An Institutional Approach to the Göta kanal

Björn Hasselgren

An Institutional Approach to the Göta kanal

A Nineteenth-Century Infrastructure Mega-Project

Björn Hasselgren
Department of Economic History
Uppsala University
Uppsala, Sweden

ISBN 978-3-031-44418-0   ISBN 978-3-031-44416-6   (eBook)
https://doi.org/10.1007/978-3-031-44416-6

Cover credit: © Melisa Hasan

This Palgrave Macmillan imprint is published by the registered company Springer Nature
Switzerland AG
The registered company address is: Gewerbestrasse 11, 6330 Cham, Switzerland

Paper in this product is recyclable.

For Marie

PREFACE

Transport infrastructure systems represent major investments signalling the long-term belief in a better future. The improved transport services that are ultimately the goal of transport infrastructure are intended to shorten transportation time, lower cost, even out social and geographical differences in nations and often to strengthen the strategic military defence. There is, however, a major difference between the necessary stages in realization of transport infrastructure systems, such as planning, construction and project management of these major projects and the following use of the systems.

The challenges related to the realization of the infrastructure are often overlooked. Over time, and still today, this can lead to long delays and to cost overrun compared to plans. One of the major aims with the current book is to shed light on one of Sweden's largest transport infrastructure projects, the Göta kanal, in order to incorporate that project into a broad co-evolutionary and institutional analytical framework presented in my 2018 book *Transport Infrastructure in Time, Scope and Scale, An Economic History and Evolutionary Perspective*. Seen together, the two books give a complementary picture of the nineteenth-century history of transport infrastructure development in Sweden.

Over time, transport infrastructure systems have been constructed even though financial resources have been short and technological and managerial skills have not been developed in full to meet the requirements of these challenging projects. Financing has been one of the challenges, and the projects have often been organized in forms where private and public financing has been combined, such as with the Göta kanal. Strong proponents of the vision to be achieved with the transport infrastructure projects have had important roles in actually making the projects come true. Baltzar von Platen, the initiator and long-standing Director of the Göta kanal corporation, was such an entrepreneur which has a major part in this book about one of Sweden's largest infrastructure projects.

This book is the result of research started in 2017. Some of the chapters have been presented as drafts in seminars at the Department of Economic History, Uppsala University. In other cases, earlier versions of texts have been presented in research conferences such as the World Economic History Conference, the Swedish Historians meetings and the Swedish Economic History Meeting. Comments on drafts for these presentations have been received from Dan Bogart and Göran Ulväng. Jan Ottosson at the Department of Economic History, Uppsala University, has given valuable comments through the project kindly sharing his profound knowledge of the transport infrastructure area.

The study concerning the relations between the iron industry at Stjernsund and Göta kanal was facilitated by a grant from the Royal Swedish Academy of Letters, History and Antiquities (Kungliga Vitterhetsakademien), Dr Carl Kempe's Foundation. Elisabet Lagerström at Stjernsund, Gustaf Svensson at Skyllbergs Bruks AB, Arvid Jakobsson at the Royal Bernadotte archives, Mats Hemström at the (Stockholm) Royal Palace Archives and Carl Gustaf Burén have all been very helpful. Roger Altsäter CEO at AB Göta kanalbolag has been very supportive through the entire research project.

Skillinge, Sweden
October 2023

Björn Hasselgren

CONTENTS

LIST OF FIGURES

LIST OF TABLES

CHAPTER 1

The Development of Transport Infrastructure Systems

Transport infrastructure systems and the development of these over time is the focus of this book. In a recent book (Hasselgren, 2018), I have presented the more general view on these systems and their development. In that analysis, the focus was roads and railroads, which are the systems that dominate the current set of transport infrastructure systems.

Here the aim is to add another transport infrastructure system to the analysis: Canals. The canal in centre of the analysis is the Swedish Göta kanal project during the preparation and construction phase from 1800 to 1832. This purpose opens for a focus on the early modern age and more specifically the time-period when Western economies' industrialization transferred from Britain to other economies, in this case Sweden. Comparisons are made with two contemporary major canal projects; the Erie Canal in the US (1817–1825) and the Caledonian Canal in Britain (1804–1822).[1]

Earlier versions of parts of this chapter were presented at the XVIII World Economic History Conference, Boston, USA, 2018. Comments were received by Dan Bogart. Drafts of chapters were also presented at the Uppsala University Economic History Department's seminar series in 2018 and 2019, at the 2019 Swedish Historians' meeting and the 2021 Swedish Economic History Meeting.

B. Hasselgren, *An Institutional Approach to the Göta kanal*, https://doi.org/10.1007/978-3-031-44416-6_1

Inland waterways and inland navigation have been used for a long time in most economies and countries where rivers and lakes have made the navigation by different kinds of boats and vessels possible. Diverse ways of overcoming obstacles to the efficient use of these inland waterways have also been in use since long. Banks in rivers hindering boats passing have been excavated and sections of rivers and lakes too narrow or with insufficient depth have been widened and deepened, often with state support.

Road transport has always been the alternative to transport with different maritime vessels. Roads have been constructed since pre-modern ages, with different technologies and capacity. Well-designed roads with a specific road structure as a basis for a stronger surface material as stone or gravel have made also heavy transport possible. Roman roads are of course one such example.

The general situation as regards transport systems in most Western economies by the late eighteenth century was that roads were insufficiently provided; both as regards the capacity in general with less appropriate surface materials and the general construction standards. Roads were too narrow, too steep and too badly maintained to fulfil the need for the growing demand for transportation in economies where trade was increasing in parallel with accelerating growth.

Here inland waterways and canal systems offered a solution. With the construction of navigable waterways without waterfalls/currents as hindrances, as in most rivers, a more reliable heavy weight transportation facility was provided, primarily for freight transport, but also for passenger transport. Compared to road transport on less well-functioning roads, canal transportation was energy-efficient and relatively fast. Indeed, the possible energy savings from a well-functioning canal transportation system was among the most used arguments in favour of canal proposals, likewise in most countries.

The considerable number of horses that could be saved, should a canal operation be realized, would reduce the general need for horses leading to a number of savings in the economy, horse breeding, distribution of fodder, etc. Calculations of the number of horses to be saved were referred to by canal promoters. On the negative side of the calculation were high investment costs, lack of technology and sometimes a insufficient institutional setting. A major drawback with inland waterways in northern geographies like Sweden is of course also the freezing of the canals during winter months, which enabled 6–8 months of operation at the most.

From a modern perspective, it is crucial to put the canal technology of the late 1700s and early 1800s within the correct frames. Canal transportation systems were, from a modern standpoint, very slow and had very limited transportation capacity. The canals, even the major ones, were in general ca 7–12 metres wide and 2–3 metres deep. The locks, which define the capacity limitations in most cases were rarely more than 30 metres long. Canal boats had to be towed from the side of the canal from a towpath, where horses or oxen (or men) dragged the boats. Some of the canals might be seen as systems of barges more than a transport system open for all boats. See, for example, the discussion on the dimension of canals in France as compared to Britain referred to by Geiger (1994) discussing the French canal programme of the first decades of the nineteenth century.

Sailing was an option in some cases, primarily where lakes allowing for sailing ships connected to the canal system, as for Göta kanal and the Caledonian Canal. Steam ships were introduced more generally from the 1830s, something which changed the capacity of canals and strengthened their competitive edge in relation to road and railroad systems. But this later development is outside the scope of this book. The motorization of canal transportation is related to a period where railroad-, and later road-transport, competition defined the continued development of the canal systems. Transportation and transport infrastructure systems have in general been developed both in competition and in parallel.

The introduction of a new mode of transport or a technology shift for one transport mode thus often is met by technology development also for other modes. Transport modes who function in competition though also often work in coordinated ways, strengthening each other. Early railroads thus worked as feeder systems to canals while later the situation was more or less the opposite. Also, today similar development paths are visible with electrification challenging the fossil fuel market for road transportation, with different strategies for response seen among fossil fuel providers and vehicle manufacturers.

THEORETICAL PERSPECTIVES

A general Analytical Framework

Transport infrastructure systems are complex and functioning within a multifaceted setting where, technical, economic, social and political systems coexist and co-evolve. They can be described, understood and analysed from several different perspectives, including many different scholarly fields. The analysis in this book is framed in an institutional evolutionary theoretical view, as presented in my 2018 book *Transport Infrastructure in Time, Scope and Scale*. Thus, the Göta kanal project, and the canal era, is seen from *a technological, an economic and a political* perspective as depicted in Fig. 1.1. This conceptual theoretical framework is used in the book in order to structure the discussion on the Göta kanal project over time.

The categorization of the shifting focus of arguments and situations over time related to the distinct categories in the analytical framework is one of the uses of the multifaceted institutional view in this book. Over time the interdependence of the varied factors influencing the development leads to a shifting balance of the basic responsibility for the projects on the public and the private sectors. The interdependence between the public and the private sectors often explains the development of large

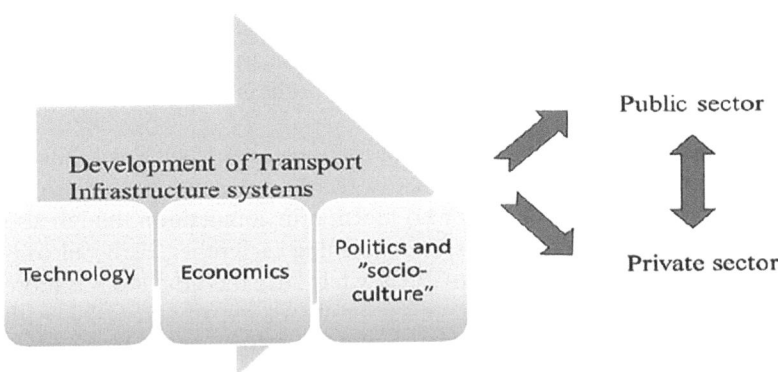

Fig. 1.1 Development of transport infrastructure—as dependent on three areas of influence leading to public vs private sector organizations (*Source* Hasselgren [2018])

infrastructure projects like canals and have been the theme of lengthy debates and disputes, but also for the development of economic theory in general, for example, as regards concepts like natural monopolies, as exemplified by Mosca (2008). Questions in relation to business or social cost perspectives, phenomena like externalities and monopoly tendencies and many more are linked to the construction and use of transport infrastructure assets.

Transport infrastructure is among the most challenging investment projects that actors like governments embark upon over time. Many of the conflicts and disputes between stakeholders have been illustrated and played out in relation to the different infrastructure projects and the specific circumstances at various times in history. The history of nation-building is connected to transport infrastructure, either for the development of the economy and growth in general or for the transportation sector as such in a narrower perspective. Important business sectors have their origin in and relation to the transport infrastructure sector, such as the shipping industry, railroad industry and vehicle/car industry.

Transport infrastructure has also been a basis for the development of organizational and financing solutions. Large managerial and communicative challenges related to new transport modes made new organizations necessary for canals, railroads and roads. Chandler (1977) even argues that the organizational capabilities and the more integrated organizational structure of railways was an important aspect of why the railway companies' outcompeted canals.

Long-term investment needs in relation to transport infrastructure supported the development of new financial instruments like specific infrastructure bonds or other financial innovation, in the case of Göta kanal in the form of a bank connected to the canal corporation. New charging and financing methods had to be developed to raise cash flow for the funding of debt. Such fee-financed transport infrastructure assets were to be found in many countries and in most transportation modes up to the twentieth century; Coase's example with lighthouses financed by private sector actors is one example; turnpikes in the US and in Britain another; canals and railroads paid for by fees/tickets a third. Only during the last 100 years have financing from governments increased to become a dominating source of revenues for the transport infrastructure systems in most countries and cases.

Göta kanal and the Analytical Framework

The Göta kanal project displays many aspects of transport infrastructure development which can be structured according to the co-evolutionary institutional development framework above. Novel *technology* and technological innovations made the construction of advanced structures like canals more easily realized with new technologies introduced during the early 1800s. Modern construction technology (railways for the transportation of earth and stone, etc.) and the use of iron-supported locks and bridges were important aspects of kanal construction. The major canal projects were also drivers of technology development. In the case of Göta kanal, this is clear with the establishment of the Motala verkstad in 1822 as an auxiliary to the canal corporation, and one of the first to adopt new production technologies.

Economics and *organization* clearly had important roles in the planning and realization of the project. Who should finance the project and how should it be funded? Should the canal be owned by the government or by a private sector corporation and who should be able to bear the different risks of the project? Should the canal project be treated as business operation merely functioning based on a narrow financial business perspective or should a broader socio-economic perspective also be part of the projections of the effects of the canal? These were themes heavily discussed for Göta kanal and for other canals under discussion and construction in Sweden and elsewhere at the time. It also turned out to be among the most challenging and important aspects influencing the Göta kanal project.

Politics and political considerations were indeed involved in the Göta kanal project. It was initiated by the pre-revolutionary regime based on a royal decree in 1808 but took off following the revolution and liberal restoration in 1809 with final decisions taken in 1810. During the project, various aspects of the construction work and realization project were themes coming back in the public debate and in the sessions in the Parliament (Riksdagen). Not until the canal was finished did the political debate related to the canal project (more or less) end. For the opposition to the government and the new regime, represented by Crown Prince Karl Johan, criticizing the Göta kanal project could be used an effective and rather safe way of keeping the government under constant scrutiny from the Parliament, without the risk of being met by oppression from the government.

Different Perspectives on Canal Projects—And Other Large Infrastructure Projects

Following a more traditional view of canal projects at large, and with Göta kanal as an example, focus has often been put on the importance of a strong forerunner, or entrepreneur, furthering the project. The project has often been seen as completely dependent on the strong entrepreneur, without whom the project would not have been initiated, support won for the idea, nor the construction being finalized. In the case of Göta canal, the strong forerunner that has been focused has with all right been Baltzar von Platen, who (re-)initiated the idea for the Göta canal to be realized and led the works and the Göta canal company up until his death in 1829. This holds for many of the British canal projects connected to for example, Thomas Telford (Caledonian Canal) or James Brindley (Bridgewater Canal), but also for the New York State Governor De Witt Clinton, the most obvious forerunner over time for Erie Canal.

The focus on single entrepreneurs and forerunners for the canal projects can be seen as part of an era of hailing these historical figures, exemplified with a number of innovators and technicians in Britain during the eighteenth and nineteenth centuries, as described by MacLeod (2010). The commemoration of these historical and even contemporary historical figures was something that became connected to the production of national stories told about the initiation of the industrial revolution and modernization of society in general. The production of statues, streets and schools named after the "hero of invention", books, paintings and museums, etc. seems to follow a general pattern in many countries in the early stages of industrialization. Göta kanal and its connection to Baltzar von Platen is indeed one of these examples.

The broader view on transport infrastructure investments and projects applied in this book does take the importance of strong forerunners or entrepreneurs into consideration, these are indeed important for the understanding of the projects. At the same time, it is crucial to see the canal projects and the "canal-mania"-periods in different countries as a result of many different processes and circumstances in the economy and the society at large. With this broader perspective canals, as a societal phenomenon, can be more thoroughly understood and placed within a framework of general development in the respective economies under study.

Transport infrastructure projects such as canals include many areas of scholarly and professional discourse. They therefore inform us both of the complexity of such projects and processes at large, but also give us a good understanding of the development of many of the issues still very much in the forefront of transport infrastructure-related discourse. In this way, studies of eighteenth- or nineteenth-century canal projects have a bearing also in our time. Many of the dilemmas facing canal constructors in earlier centuries are also relevant today.

A transport infrastructure project like a canal is generally structured, or afterwards possible to structure, as a process spanning from idea generation, through planning and support generation, to decision-making, financing and organization issues, further to realization, with questions around the effective design of the project, the relation to contractors or acquiring of resources in other ways, the public debate often surrounding projects like this and finally to completion and subsequent use of the asset over time. Structured in this way, we clearly see the diversity of scholarly fields that come into play when transport infrastructure is analysed.

The following fields of theorizing and research can be seen as involved in the various stages of a project:

- Initiation—entrepreneurship and political economy/bargaining procedures in a public sector setting involving game theory, capabilities to build support and to reach out to the public, where both financial resources, social status and rhetoric strength counts.
- Planning—planning theory and the management of uncertainty and risk—what kind of documentation is necessary to produce at what time and with what quality, to control risk and cost.
- Design—questions relating to how to design the infrastructure—use of well-known technologies and solutions or the introduction of novel technologies and innovation at large.
- Financing and organization—should the project be seen as a "standalone" solution and thus a separate entity owned by private owners serving primarily a commercial interest or as a public sector interest with public sector financing and ownership? What kind of support is necessary to create taken the two different organization models? Theories from economics, business management, public choice and political science come into play.

- The construction—how to organize the construction phase— with own resources (management, workers, equipment, etc.) or with contracted construction corporations—questions in relation to procurement, principal/agent-theory, contract theory and construction management theory are involved in this part. Risk management is an inherent discussion. Who bears the responsibility for risk in different phases of the project?
- Completion—as the infrastructure is opened for use, new questions arise: how will it be managed and how will prices and fees be set? Should the canal be seen as self-sufficient covering all its costs or managed as a public sector entity where fees should be set in relation to some kind of social-economic principle? How to manage questions in relation to monopoly and competition? Are there political interests that influence how the canals are used and priced? The competition with other transport modes, such as railroads. Are the canals outcompeted by railroads or do they support each other during a process of intermodal shifts and competition? Another question is of course how important the different transport modes are for growth. Did the railroads bring added growth or was the net benefit to the economy relatively low, compared to canals?

The wide range of questions that come into play when large infrastructure projects like canals are analysed often make these projects difficult to manage for owners, users and political systems. They seem to get a life of their own, difficult to stop once they have been initiated and difficult to control. Altshuler & Luberoff (2004) have discussed these projects as *Mega-projects,* where many of the aspects outlined above are included in the discussion in a contemporary setting. The tendency for such mega-projects to be difficult to manage in the political system having a life of their own has been analysed by Flyvbjerg and others, for example, under the heading of *Rationality and power* (Flyvbjerg et al., 1998). In this book, these perspectives form part of the general theoretical background for the discussion of Göta kanal and the other canals which are brought into the discussion as comparisons.

The wide range of theoretical fields that can be applied to the analysis are difficult to cover in their entirety. Some of the major strands of these will be touched upon in the following chapters. Since this is a discussion focusing on the early 1800s, an economic history perspective is though the basis for the use of the theories. A canal project can thus be seen as a mega-project in the sense we discuss these today, the reflections on theories and scholarly discourse will though also have to be based in the thinking and views of the early 1800s in order not to lose in relevance. More on this below.

NOTE

1. Britain has been used for England and Great Britain.

REFERENCES

Altshuler, A. A., & Luberoff, D. E. (2004). *Mega-projects: The changing politics of urban public investment*. Brookings Institution Press.

Chandler, A. D. (1977). *The visible hand, The Managerial Revolution in American Business*. Harvard University Press.

Flyvbjerg, B. (1998). *Rationality and power: Democracy in practice*. University of Chicago Press.

Geiger, R. G. (1994). *Planning the French canals: Bureaucracy, politics, and enterprise under the restoration*. University of Delaware Press.

Hasselgren, B. (2018). *Transport infrastructure in time, scope and scale, an economic history and evolutionary perspective*. Palgrave Macmillan.

MacLeod, C. (2010, April). *Heroes of invention*. Cambridge Books, Cambridge University Press. ISBN 9780521153829.

Mosca, M. (2008). On the origins of the concept of natural monopoly: Economies of scale and competition. *The European Journal of the History of Economic Thought, 15*(2), 317–353.

CHAPTER 2

Göta kanal and the "Canal Age": The Literature

The Göta kanal project has generated a vast literature in Sweden (and predominantly in Swedish) since the nineteenth century.[1] Only few scholarly texts in English have been published with the canal project as their core subject. The current study thus fills a gap in the literature, where Göta kanal and the experiences from it can be inserted into the general framework of inland waterways studies in other countries and major infrastructure projects in general. Thus, analysing the Göta kanal project in relation to an institutional co-evolutionary analytical framework and offering improved possibilities to compare it to other projects is among the aims of the current book.

Looking into available sources to Göta kanal, there is a vast archive material on the canal, both the AB Göta kanalbolag's (the canal corporation formed in 1810) office in Motala and the archives at the National Swedish Archives in Stockholm (Marieberg and Arninge) and in Vadstena. These archives (Arninge/Marieberg and Vadstena) covering general Göta kanal-related documents and specific Göta kanal Diskont (the bank-institution originally connected to the canal company according to the 1810 Privilege decision) documentation have been studied on a general level as a preparation for this book.

Göta kanal has been the theme for numerous discussions, proposals by the government and single members of Parliament, public investigations

© The Author(s), under exclusive license to Springer Nature Switzerland AG 2023
B. Hasselgren, *An Institutional Approach to the Göta kanal*,
https://doi.org/10.1007/978-3-031-44416-6_2

and decisions over time. Many of these documents are available at the National Archives, but also at the Swedish Riksdag Library, Stockholm. Even though the documentation in these archives and libraries is very extensive, much additional information in relation to Göta kanal is available in archives of the major players like Baltzar von Platen, who initiated the project and led it until his death in 1829, and other prominent actors like the Swedish cabinet members at the time, who were in more or less constant contact with von Platen during the project. The archives with letters and other documents by Baltzar von Platen are kept, to a considerable extent, at the Royal Library in Stockholm, and these have been studied with a focus on Göta kanal-related documents. Von Platen's contacts and collaboration with one of the other centrally placed players in the 1809 events in Sweden, Georg Adlersparre, can be studied in the Lund University Libraries, Sweden. Documents from these archives are widely cited in the literature.

One of the major primary sources in the literature regarding Göta kanal is Samuel Bring's (et al.) *Göta kanals Historia* (1922/1930), originally published for the centenary of the inauguration of the first stretch of the canal, which gives a thorough and broad description of many of the different aspects of the Göta kanal project, but primarily with a Swedish perspective and background and lacking a specific theoretical framework. Though being one of the most comprehensive descriptions and being based on the original archival documents in relation to Göta kanal, the text unfortunately lacks explicit references, for example, to the, often lengthy, excerpts from these documents. This is also the case for the references to Parliamentary debates from the four decades of the canal project studied in this book. The value of Bring's major work as a more specific source for research is, therefore, unfortunately limited. In general, though, it offers a good overview of the Göta kanal project and is used in this book as a source in those cases relevant references are given.

Two PhD dissertations were published (in Swedish) during the 1990s at Linköping University, the Swedish university found closest to (anyway to the eastern section of) Göta kanal. Lars Strömbäck in his (1993) *Baltzar von Platen, Thomas Telford och Göta kanal. Entreprenörskap i brytningstid*, focuses on the close cooperation between the canal project in Sweden and the British canal projects at the time. The close collaboration between Baltzar von Platen and Thomas Telford is a central theme in this dissertation. The British projects, with a direct parallel to the Scottish Caledonian Canal are discussed to some extent by Stömbäck. Here, some references to the British canal projects are given and a brief discussion on other contemporary projects is included.

The Göta kanal project is also analysed by Strömbäck as an example of *Dahménian Development Blocks* theory (Dahmén, 1988), where the close collaboration between the Government, the project and technology providers at the time is a theme. This perspective comes close to the analytical framework in this book.

Hans Lindgren (1993) offers a historical perspective in his 1993 dissertation *Kanalbyggarna och staten, offentliga vattenbyggnadsföretag i Sverige från medeltiden till 1810*, where the long history of inland waterways planning and construction in Sweden is outlined and analysed. Here the close connection between the preceding Göta Älv and Trollhätte kanal projects is one focus, among many other canal projects since the medieval times in Sweden. The balance between government and private sector initiatives, the latter of which there has been a number over time in Sweden in relation to canals and other inland waterways, is discussed.

Lindgren's theoretical basis is something like a mix between a resource-based view relating the capabilities and available resources of the central actors in canal projects to the institutional setting at the time of different canal projects. Lindgren though explicitly treats technology as an external aspect to the understanding of canal projects. He argues that the construction methods remained unchanged during the centuries before the Göta kanal project. This seems, obviously, as a less fruitful theoretical framework for an analysis of Göta kanal, a project which to such a high extent was involved in introducing new technologies both for the construction of the canal and when it comes to the construction of locks, etc. Lindgren's interest for organizational aspects like the balance between the king/government and private innovators/entrepreneurs though comes close to the perspective used in this book.

Lindgren's approach is important also in that it brings into the discussion other Swedish canal projects, before and after Göta kanal.

Lindgren shows how these canal project (and more) were planned and realized, often with a close collaboration between the government and private entrepreneurs and/or companies set up. *Strömsholms kanal*, a late eighteenth-century Swedish canal project connecting the iron industry region north of Lake Mälaren to the lake, is a canal project which is similar to Göta kanal in this organizational aspect (a separate corporation set up and considerable financing raised from private sector owners), but is closer connected to a specific industry sector, the iron industry, compared to Göta kanal with its broader general commercial outlook.

A good description of Strömsholms kanal and the construction process is given by Schenström (1797), one of the leading actors in the project, in a contemporary thesis. A thorough presentation of the financing arrangements is part of Schenström's thesis. Also, in this case there was a shared private and public sector involvement in the financing of the canal. There was, indeed, a good deal of experience from organizing canal projects and from canal construction at the time of Göta kanal. None of the earlier canals, though had a similar length, width or depth. Göta kanal was unprecedented in Sweden in its grand scale.

A recent doctoral thesis by the French historian Gauchet (2020) is also devoted to Göta kanal and during the same time as reflected in this book, 1800–1832. In the thesis, the perspective is primarily political, but the project is also viewed from a technological, financial and power perspective. Gauchet is arguing that the project was part of or developed into a part of the regime of Karl Johan to gain power over and to further the construction of inland waterways, harbours and engineering and engineering education in general.

While offering a broad and well-researched version of the important events during the construction of Göta kanal, the latter purpose to show how Göta kanal was used in the broader sense for the government's wider political aspiration in the field of transport is though perhaps less well justified by the sources and facts Gauchet presents. The thesis is though important as a reminder of the European-continental and in specific French view on Sweden and its early nineteenth-century political history rather than offering an alternative understanding of the Göta kanal project.[2]

Some of the leading actors in the Göta kanal project issued books and reports as part of the contemporary discussions in relation to the canal, and some of these actors have also been the subject of later biographers.

Baltzar von Platen, the leading figure in relation to the project, was himself a diligent author of writings about Göta kanal. Among the more important sources to the discussion during this time are his writings from the period 1806–1822, which argue for the project, describe how it can be implemented in detail and finally argue that the chosen form of implementation is the right one and why (von Platen, 1806, 1822; von Platen & Telford, 1809). In these books, or as we should say today pre-studies, project plans, monitoring and evaluation reports, von Platen has different roles, at first in 1806 in a role proposing the project and arguing for its benefits, and relatively modest calculated cost.

In the 1809 report, the style is more specific where more detailed aspects of the project are described, following the work during the summer of 1808 together with Thomas Telford, who visited Sweden and eventually stayed in von Platen's home, Frugården. The 1822 report is a monitoring and evaluation aspiring report going through the project up to this time reflecting on the measures taken, the organization and cost so far of the project. The project is also compared to other contemporary projects with the purpose of justifying the development of the project and describing why costs have soared in relation to the original budget calculations, etc.

Von Platen held a number of offices besides the roles as managing director and chairman of the board of Göta kanal corporation. In relation to many of these offices, von Platen published reports and other statements in a broad variety of subjects, primarily though in relation to naval themes and questions in relation to Sweden's union with Norway, where von Platen served as the King's governor to Norway from 1827 until his death in 1829. There is no recent or contemporary biography on von Platen, besides more popular books reflecting on von Platen in relation primarily to Göta kanal.

The former prime minister of Sweden Louis de Geer (1887) published the primary more serious biography of von Platen by the end of the nineteenth century. In comparison with the numerous books that have been published about Thomas Telford and De Witt Clinton, this is surprising, considering the extent to which von Platen has been commemorated in public writing in relation to Göta kanal and other circumstances, such as naming streets after him in many different places in Sweden. This outstanding enduring figure in the history of Swedish technology introduction and innovation therefore is surprisingly anonymous at the same time as clearly present.

Some of the other actors in relation to Göta kanal have been focused in recent biographies. Hans Järta, the state secretary for financial matters, at the time of the establishment of Göta kanal corporation and financial support to the project, is the theme of a book by Petri (2017), where the close relation between von Platen and Järta during the 1809 events comes to the fore, as does the shared interest for a close relation to Britain.

One of the important actors in the relation between the King, the Cabinet and the Göta kanal project was Carl David Skogman, one of the directors of the Göta kanal Diskont (the bank connected to the project) and later finance minister of Sweden, succeeding Järta some years later. A biography of Skogman is given by Körberg (2009). The book gives some important background information as regards the remarkably close relations between the King, the Cabinet and the Göta kanal corporation. Skogman was interested in classical-liberal thoughts and ideas and read Adam Smith's works early on. He also travelled both to Britain and the US during the early years of the 1800s on behalf of the government to gather information on the ongoing economic development.

The financing and financial aspects of the Göta kanal project has been studied by Kärrlander (2011). Kärrlander compares Sweden's three early modern banks in operation during the first two decades of the nineteenth century. One of these was the "Göta kanal-Diskont", even if the primary focus in the dissertation is the 1817 crisis of the Malmoe-situated sister bank of the Göta kanal bank. A further background to the financial arrangements and organizations from the late 1700s but including also the Göta kanal Diskont is given by Fritz (1990).

Regarding the structure and changes in the transport market in Sweden during the period studied, Thomas Thorburn's (2000) very ambitious work, which describes the period 1780–1980, is extremely helpful. The long-term development of transport flows in Sweden and the market shares for different modes and goods are well-described. It shows, for example, the market share for inland waterways compared to land transport and railroads.

Another archival resource which brings light to the Göta kanal project is the Thomas Telford archive at the Library of the Institution of Chartered Engineers (ICE) in London, UK, where letters sent to and received from von Platen are kept. Telford was one of the central figures in the nineteenth century primarily British infrastructure and built environment-era acting as chief engineer in many projects similar to the Göta kanal. Telford also was the initiator of the ICE, which is still today a leading professional institution for engineers with a global outreach. The Telford archive and collection of letters is in an excellent condition.

The work of Thomas Telford is put into perspective in a 1959 biography by Rolt (1959), wherein the projects Telford engaged in and the engineering skills that was a prerequisite for a number of grand infrastructure projects in Britain and other countries where Telford contributed are described. Canals, bridges and tunnels where among Telford's main achievements, but Telford was also involved in road construction and a critical voice in the debate around the introduction of railroads. The Caledonian canal project is given a thorough description in this book and is complemented by Burton's *The Canal Builders—The Men who constructed Britain's Canals* (1972/2015). These two books can be seen as examples of a history writing tradition which puts the main emphasis on the men who figured in the forefront of the projects, but with lesser interest and focus as regards the broader context.

The actual organization of canal construction projects in Britain during the canal era is described by Cross-Rudkin (2010). Here some of the main players in the growing construction business developing as a result, partly of the canal era is described. The situation in Britain in this respect certainly contrasts against the rather undeveloped situation in Sweden at the time.

The Erie Canal has been extensively researched in different scholarly traditions. A major review and tale of the origins and effects of the Erie Canal is given in Ronald E. Shaw's (1966) *Erie Water West*, which brings the canal into the background of technological, economic and political developments primarily in a New York State perspective. Shaw shows how the canal had a path-breaking role in the development of New York State, its trade patterns in general and for politics during a number of decades. A broader historical description that focuses more on social and political processes in connection with the Erie Canal construction and the upcoming process during the operational phase is described by the historian Sheriff (1997).

Competition between and co-evolution among the Erie Canal and the railroads is the focus in Ellis's (1948) *Rivalry between the New York Central and the Erie Canal*. Here the interdependence between canal construction and financing and railroad construction and policy is studied. The early development of anti-trust and anti-monopoly policy in the US is seen as integral to the growth of railroads as contrasted against the Erie Canal.

Both Ellis and Shaw emphasize the importance of transport infrastructure, and the major investment these bring, to the constitutional developments both at the State-level and for the federal political debate during the nineteenth century. The republican democratic system's way of managing these issues, with a constant public debate and tension, is thereby contrasted against the constitutional monarchies Britain and Sweden, where other sources of political power come into the fore and a more moderate political discourse was developed with the monarch as a factor both in a role to drive the development and to moderate.

The response to the construction of Erie Canal from other US cities on the east-coast is the theme in Rubin's extensive 1961 article (Rubin, 1961). Rubin compares the response of Boston, Philadelphia and Baltimore to the favoured situation of New York and Albany, following the 1825 completion of Erie canal. The main alternatives, according to Rubin being either to imitate the case of New York by constructing canal systems of their own or by going for the railway technology, by the mid-1820s a novel technology to the US. The responses turned out differently in the different settings, and reasons for that are discussed by Rubin. In the case of Sweden, it is not unlikely that a decision on Göta kanal taken by the mid-1820s instead of the early 1810s would also have played out differently. Could it perhaps be argued that the introduction of railways in Sweden was delayed by the Göta kanal project, blocking other major investment plans for a long time?

Another source of inspiration and reference is naturally the vast literature concerned with technology development and history. When it comes to the general discussion on industrial enterprise and technology, Chandler (1977) is a useful source. Evans (1981) supplies a discussion on the competition between canal and railroad technology in Britain during the first decades of the nineteenth century which is of great value to the understanding of technology development and its challenges.

Olmstead (1972) discusses Erie canal from a financing point of view. The co-evolution of savings banks in New York State with the development of the Erie canal project, a largely private sector initiative, though connected to public sector policies, stand out as similar to Göta kanal with its own bank, the Göta kanal Diskont.

Dan Bogart has highlighted several aspects of shipping and canal construction from an economic-historic point of view. An example is an article about party affiliation and faction formation in connection with the transport revolution in Britain, which shows similarities with the political discussion in Sweden around the Göta kanal project, where the adjacent towns to the east in Sweden, Söderköping and Norrköping, were for a long time in conflict over the starting point of the new canal. Söderköping "won" the struggle for Göta kanal's eastern connection to the Baltics, but Norrköping overall has experienced a stronger economic growth. Canals have indeed been the origin of numerous struggles, political and economic.

The question of the connection between infrastructure investments and the beginning and continuation of the industrial revolution has been analysed by many researchers. Trew (2010) makes a broad comparison between different infrastructure systems in the UK, where channels are included as an example, and points out that one can see the history of the development of infrastructure systems as a long-term learning process in the field of financial technologies and methods.

A classic in the field of financing canal projects is otherwise Ward (1974), who presents a very thorough review of canal projects in Britain during the eighteenth and nineteenth centuries. The focus here is on the projects themselves, the financiers, and the financial technology more in detail. The participation of different segments of society in the financing of canal corporation, as regards social strata, gender and geographical closeness to the specific canal project is explained clearly by Ward.

A similar investigation in relation to the French 1814–1848 canal era is given by Grosskreutz (1977). In this book, a very thorough analysis of the entire canal-building programme set in motion by the restoration government is at the core of the presentation. All of the canal projects are included, with a focus on the financing of the canals and to which social groups the main financiers belonged. Hereby the collaboration between the government and the dominating private banks, primarily in Paris (the "Haute Banque"), but also with local actors and foreign interests is described. The balance between the state and the private sector and the varying degree of private financing is brought to the fore.

Complementary financial perspectives on canal construction can be found in Arnold & McCartney (2011), who point out that the return seems high in several of the early canal projects in Britain, but that this can be explained by the fact that the accounting rules were used relatively freely and not with the meaning we usually see today. For example, companies were often seen from a one-year perspective with more of a liquidity perspective than a longer-term cost perspective.

As regards the ownership of the British Ellesmere and Chester canals, Wilson (1957) has presented more detailed data. These are based on lists of shareholders from the two canals found by accidence in relation to archival research. Ward has developed the analysis of the ownership into a more structured presentation with different categories of owners.

A more general perspective on the balance between public and private involvement in infrastructure systems in Europe from the nineteenth century onwards can be found in Millward's *Private and Public Enterprise in Europe, Energy, telecommunications and transport 1830–1990,* where infrastructure development with a focus on the balance between public and private financing and ownership is brought to the fore (Millward, 2005). A continuous shift between public and private sector responsibility seems to be the ruling principle in this development.

The more organizationally oriented description, mainly of the railway systems, that Chandler (1977) made in his classical analysis of management and organization also includes, less often referred to, a background from the canal-building era. Before the coming of the large railway corporations, there was an important organizational and managerial learning taking place in relation to the canal enterprises.

Nation-building as a motivation for large transport infrastructure projects but also other possible comparisons can be made between the European projects and the growth of the nineteenth-century infrastructure systems into the twentieth century. The connection between the establishment of national identity and science/technology is discussed in an anthology edited by Harrison & Johnson (2009), where channel projects such as the Erie Canal are highlighted from this perspective. This fits well into the politics and "socio-culture" explanations to the development of transport infrastructure in Fig. 1.1 in Chapter 1.

The term mega-projects in the title of this book refers to the book by Altshuler & Luberoff (2004) exploring the functioning of large transport infrastructure projects during recent decades in the US. This book reveals the multifaceted reality of such projects that often span over long-time horizons and where national and local politics is blended with technology, construction and management issues and financing. Often such projects seem to evolve into entities of their own being too big to fail in the end. As we will see, this is something that characterized the Göta kanal project over time.

This introduction to some of the literature relating canal projects to societal development at large, and to the inner life of canal projects, is of course very incomplete. It though exemplifies the large diversity of the literature and the many aspects which have been researched. It is the intention in this book to make use of the literature mentioned here when analysing the Göta kanal project in relation to other transport infrastructure projects and to other canal projects.

All of the major canal projects during the eighteenth and nineteenth centuries, which are focused here, were initiated and carried through against the background both of national or local circumstances and the amalgamated knowledge of how to execute these major projects. Some of the projects were less connected to the general development and thus more local than others, where the canal project "know how" was more put to use. In this way, the different specific projects communicate to varying degrees with other earlier, contemporary and later canal projects. The Göta kanal was initiated in a very specific challenging time for Sweden seen from a "geopolitical" point of view, with the loss of Finland to Russia in the 1809–1810 war and the 1809 revolution. Other projects have their specific motivation or background. As we understand these specificities of the different canal projects in different countries, we can also better understand the way the projects worked and were realized.

NOTES

1. An extensive bibliographical publication covering Göta kanal and other Swedish canals was issued by Linköping University in 2001 in five sections:

   Brage, C. (2001) *Göta Kanal. Forskning från Linköpings universitet; 14* (Del 1 av 5) Linköping University Electronic Press Linköping, Sweden.

   The bibliography is also available as a database, where also maps and other documentation relating to Göta kanal is available: *Göta- och Trollhätte kanals bibliografiska databas Göta- och Trollhätte kanal—LiU Electronic Press.*

2. I develop my view in relation to this in a review of Gauchet's thesis, Hasselgren (2022).

REFERENCES

ARCHIVES

Örebro. Folkrörelsearkivet [Social Movement Archives].
Vadstena. Landsarkivet. Göta kanalbolags arkiv [Göta kanal Corporation Archives].
Stockholm. Riksarkivet [Swedish National Archives].
Stockholm. Riksarkivet/Slottsarkivet [Royal Palace Archives].
Stockholm. Riksarkivet. Sjöholmsarkivet [Sjöholm Archives].

Government Publications

[Charter of the Göta kanal Corporation, 1810] Privilegium för Götha Canal Bolag, givet Stockholms Slott den 11 April 1810.

[Regulations for the Göta kanal Corporation, 11 April 1810] Reglor för det för det genom Kungl Majts nådiga privilegium af den 11 April 1810 Octroyerade Götha Canal Bolag Gifne Stockholms Slott den 11 April 1810.

Literature

Altshuler, A. A., & Luberoff, D. E. (2004). *Mega-projects: The changing politics of urban public investment*. Brookings Institution Press.

Arnold, A. J., & McCartney, S. (2011). 'Veritable gold mines before the arrival of railway competition': But did dividends signal rates of return in the English canal industry? *The Economic History Review, 64*(1), 214–236.

Bogart, D. (2016, March). Party connections, interest groups and the slow diffusion of infrastructure: Evidence form Britain's first Transport Revolution. *The Economic Journal, 128*, 541–575.

Brage, C. (2001). *Göta Kanal. Forskning från Linköpings universitet; 14* (Del 1 av 5) Linköping University Electronic Press.

Bring, S. E., et al. (1922/1930). *Göta kanals historia, del 1–2*. Almqvist & Wiksell.

Chandler, A. D. (1977). *The visible hand, The Managerial Revolution in American Business*. Harvard University Press.

Cross-Rudkin, P. (2010). Canal contractors 1760–1820. *Railway and Canal Historical Society Journal, 36*(7), 27.

Dahmén, E. (1988). 'Development blocks' in industrial economics. *Scandinavian Economic History Review, 36*(1), 3–14.

De Geer, L. (1887). *Baltzar Bogislav von Platen: Minnesteckning*. Norstedts.

Ellis, D. M. (1948). Rivalry between the New York Central and the Erie Canal. *New York History, 29*(3), 268–300.

Evans, F. T. (1981). Roads, railways, and canals: Technical choices in 19th-century Britain. *Technology and Culture, 22*(1), 1–34. https://doi.org/10.2307/3104291.

Fritz, S. (1990). Commercial banking in late eighteenth century Sweden. *Changes in two Baltic countries, Poland and Sweden in the Eighteenth century.*

Gauchet, T. (2020). *A political history of the Gotha Canal Technology, infrastructure and power in Northern Europe (1790s–1832).*

Grosskreutz, H. (1977). *Privatkapital und Kanalbau in Frankreich 1814–1848, Eine Fallstudie zur Rolle der Banken in der französischen Industrialisierung*. Duncker & Humblot.

Harrison, C. E., & Johnson, A. (2009). Introduction: Science and national identity. *Osiris*, *24*(1), 1–14. https://doi.org/10.1086/605966

Hasselgren, B. (2022). A political history of Göta Canal: Technology, infrastructure and power in northern Europe (1790–1832) by Thomas Gauchet. *Technology and culture* [Internet]. Johns Hopkins University Press, 63(4), 1192–1194.

Kärrlander, T. (2011). *Malmö diskont—en institutionell analys av en bankkris*. Doctoral dissertation, KTH Royal Institute of Technology, Stockholm.

Körberg, I. (2009). *Carl David Skogman, Den okände makthavaren*.

Lindgren, H. (1993). *Kanalbyggarna och staten, offentliga vattenbyggnadsföretag i Sverige från medeltiden till 1810*. Linköping University.

Millward, R. (2005). *Private and public enterprise in Europe, Energy, telecommunications and transport 1830–1990*. Studies in Economic History, Cambridge University Press.

Olmstead, A. L. (1972). Investment constraints and New York City mutual savings bank financing of Antebellum Development. *The Journal of Economic History*, *32*(4), 811–840.

Petri, G. (2017). *Hans Järta—en biografi*. Historiska Media.

Rolt, L. T. C. (1959). *Thomas Telford*. London Scientific Book Club.

Rubin, J. (1961). Canal or railroad? Imitation and innovation in the response to the Erie Canal in Philadelphia, Baltimore, and Boston. *Transactions of the American Philosophical Society*, *51*(7), 1–106.

Schenström, M. (1797). *Afhandling om Strömsholms canal och slusswärk, af Magnus Schenström. Upsala, tryckt hos Joh. Fr. Edman, kongl. acad. boktr. 1797*. Uppsala: (Edman).

Shaw, R. E. (1966). *Erie Water West: A history of the Erie canal 1792–1854*. The University Press of Kentucky.

Sheriff, C. (1997). *The artificial river: The Erie Canal and the paradox of progress, 1817–1862*. Macmillan.

Strömbäck, L. (1993). *Baltzar von Platen, Thomas Telford och Göta kanal. Entreprenörskap i brytningstid*. Brutus Östlings Symposion.

Thorburn, T. (2000). *Economics of transport: The Swedish case, 1780–1980* (No. 12). Almqvist & Wiksell International.

Trew, A. (2010). Infrastructure finance and industrial takeoff in England. *Journal of Money, Credit and Banking*, *42*(6). The Ohio State University.

Ward, J. R. (1974). *The finance of canal building in eighteenth century England*. Oxford University Press.

Wilson, E. A. (1957). Proprietors of the Ellesmere and Chester Canal Company, 1822. *The Journal of Transport History*, *1*, 52–54.

Von Platen, B. (1806). *Afhandling om Canaler genom Sverige med särskildt avseende å Wenerns sammanbindande med Östersjön*.

Von Platen, B. (1822). *Försök till utredning af följderna utaf den arbets-methode vid Götha Canal blifvit brukad samt dervid använd kostnad jemte en blick på denna canals blifvande nytta.* Linköping.

Von Platen, B., & Telford, T. (1809) *Berättelser samt kostnads-och ersättnings-förslag rörande den föreslagna Göta kanalen.*

CHAPTER 3

Göta kanal: The Major Swedish Canal—And an Example of the Canal Era—A Brief Introduction

GÖTA KANAL

Proposals for making way for an inland navigable waterway between the North Sea and the Baltic Sea (from Gothenburg in the west to Söderköping in the east) had been made on numerous occasions in Sweden. A decision to build the canal had even been taken by the Swedish Parliament (the Riksdag) in 1772 (actually one of several decisions regarding a navigable connection in this relation taken by the Riksdag), but never came to an implementation. Reasons for this were lack of funds, further discussions between the towns in the east (Linköping and Söderköping), one of which would be passed by, and considerations on which would be the best routing of the canal to choose. Even if decisions were taken, there was in the end a lack of support to actually implement the decision.

The construction of Göta kanal starting in 1810 and opened in 1832 is a story of unprecedented challenges as regards planning, organization, financing and construction in Sweden. Compared to the original budget, the final cost soared 12 times, according to Bring (1930), from ca. 830.000 Rdr[1] to ca. 10.400.000 Rdr. The eventual construction phase lasted 22 years, compared to the originally estimated eight to ten years. While the Göta kanal project involved a high degree of innovation in many respects—organizationally, financially and technology-wise—the

B. Hasselgren, *An Institutional Approach to the Göta kanal*, https://doi.org/10.1007/978-3-031-44416-6_3

25

project also brought about a continuous political debate and difficulties endangering the entire project on revolving occasions.

Göta kanal connects two of Sweden's major lakes Vänern and Vättern and passes a number of lakes connecting Lake Vänern to the Baltics. Vänern is connected to the North Sea via the major river Göta Älv, which runs from Trollhättan at the southern part of Vänern to Gothenburg.

The passage from Vänern to the river Göta Älv is very steep with major waterfalls, which made the passage unnavigable until 1800, when a set of locks were put in use, a project that had been ongoing since the early decades of the 1700s, with numerous less successful attempts and different construction proposals tried. As the Trollhättan falls were overcome by a canal and the set of locks, the plans for building the rest of the earlier-decided Göta kanal were again taken up for consideration.

Göta kanal has a total length of 190 km, of which 87 km cut and/or blasted. The canal has 58 locks and the highest point is ca 92 metres above the lowest point, which is the sea level. In combination with Göta Älv and the Trollhätte kanal, the total length of the waterway is 390 km. From Lake Vänern in the west to the Baltic Sea, six lakes are connected with the canal: Vänern, Viken, Vättern, Boren, Roxen and Asplången. Vänern is the largest lake in Sweden, and Vättern is the third largest lake.

As mentioned above, there were a number of Swedish canal projects before and after Göta kanal. *Trollhätte kanal* has already been mentioned, as a direct forerunner to Göta kanal and an important point of departure for the project. Other canals were planned to connect to Göta kanal, and some of these were also constructed. *Kinda kanal*, which passes one of the major towns, Linköping, situated at Lake Roxen in the east section of Göta kanal, is a canal system (built in the 1860s) which connected inland waterways in a north–south direction to Göta kanal. *Dalslands kanal* (built in the 1860s) connects to Lake Vänern in a north-east direction, and had a role also in the relation to Swedish transport flows to Norway, as from 1814 in union with Sweden.

A number of canals were constructed that connected to Lake Mälaren, which is not directly connected to the Göta kanal system. *Strömsholms kanal* (built 1777–1795) connected the important iron industry region north of Lake Mälaren, and *Hjälmare kanal* (built in the 1620–1630s) connected Lake Mälaren and Lake Hjälmaren, which facilitated transport from the mid-Sweden region surrounding a number of medium-sized towns with each other and to the important Stockholm market. *Södertälje kanal* (built 1806–1819) made the passage from the Baltic

Sea to Mälaren navigable, offering a shorter route to Lake Mälaren and the Baltics as well as to Stockholm, compared to passage through the archipelago east of Stockholm.

Most of these canals have subsequently been renovated, reconstructed and altered in different projects. It is also worth mentioning that most of the canal projects were organized as private organizations with different kinds of government support. Göta kanal thus was part of a broader tradition for how to organize and finance canal infrastructure projects in Sweden (Table 3.1).

In the Nordic countries, the *Saima Canal*, on the border of Finland and Russia, is another example of canal-building, finished in 1856 and still in operation. The length of the canal is 43 km with a height difference of 76 m and with eight locks. Both Norway and Denmark have a large number of canals of which no other has a similar length to Göta kanal. From later date stems the German *Kiel Canal*, which was opened in the late 1800s. It is still in commercial use, connecting the North Sea to the southern Baltic Sea.

The US Erie Canal, opened in 1825, has nearly the double elevation (ca 172 metres) compared to Göta kanal and an original length of 584 km, with 36 locks. The Caledonian Canal in Scotland, built as a "twin" project to Göta kanal (with an active import of knowledge from Britain to Sweden) with the construction started in 1803 and finalized in 1822. The Caledonian Canal is ca 97 km in length, of which a third is man-made. There are 29 locks in the Caledonian Canal. The elevation is 32 metres. The capacity (width and depth) of the Erie Canal was originally less than that of the Göta kanal, which in turn had a lower capacity compared to the Caledonian Canal.

Göta kanal is a good example of an infrastructure project that is more easily understood and analysed in a fairly broad co-evolutionary and institutional framework, more than as, primarily, a more narrowly defined construction project. Such a wider analytical framework has been used in a number of studies in relation to transport infrastructure and has been explored in my book on the general development of roads and railroads in Sweden, Hasselgren (2018). In this book, different perspectives are applied to the long-term development of transport infrastructure systems to reflect upon and analyse the different aspects and factors having an influence on the development of these over time. Compared to the standard description of the Göta kanal project given, for example, by Bring,

Table 3.1 Timeline of major events of the Göta kanal project 1800–1832

1800	1806	1809–1810	1812–1814	1817–1818	1822	1829	1832
Trollhätte kanal completed	First proposal presented by von Platen	Second proposal presented and Parliamentary decision to initiate the project	First financing crisis, external construction staff hired	Second financing crisis Göta kanal Diskont dismantled, increasing government financing	Western stretch opened	Von Platen dies, final Parliamentary financing decision	Entire canal opened

this approach widens and deepens the understanding of the project as a social, economic and technological phenomenon.

In Sweden, Göta kanal still stands out as one of the very largest transport infrastructure projects to date, comparable to the railway system and the intense construction of roads during the 1950s–1980s. The project was initiated during a period in Sweden marked by strong challenge and change to the country. Sweden had been involved in a number of wars with Russia since the early 1700s and successively lost most of the eastern and southern provinces. Finland, which had been part of Sweden since the 1200s, though remained at the time when the Göta kanal project was initiated by the early 1800s.

Following the coup d'état in 1772 and the assassination of King Gustav III in 1792, Sweden had fallen into a period of political instability, varying between liberal policies and a harsh autocratic regime under the supreme king Gustav IV, the son of Gustav III who had come into power as of his 18th birthday in 1797. Though having an agenda of restoring stability in the country's finances and implementing reforms in different parts of the economy and the legal system (see Carlsson, 1944), the King lacked any of his father's strongly developed political abilities. And he was successively more and more isolated from the nobility, who held many of the important roles and offices in the pre-industrial era.

According to the Constitution (1792 Act of Union and Security), the King had the sole right to call sessions of the Riksdag. Following the Riksdag in 1800, which, according to the King's view was a disaster, airing the opposition towards the regime, Gustav IV refused to call another parliamentary session.

The foreign policy situation became extremely troublesome during the first years of the 1800s. Sweden was in the midst of the disputes between Russia, Britain and France. Successively, the relation towards Russia deteriorated and war broke out in 1808 on the eastern border of Finland. The moral of the troops was at a historic low, with open opposition towards the King, who handled the situation very un-tactical. The war was fought without any reasonable plan and with Sweden's finances once again deteriorated, partly as a result of the King's refusal to call a parliamentary session to discuss finances and taxes. The outcome was the loss of Finland and Åland to Russia in 1809 and the successive dethronement of Gustav IV in March 1809.

In the midst of this situation, the Göta kanal proposal was investigated and put forward for the first time since the earlier parliamentary decision in 1772 to build the canal. At first, a proposal was published in a report in 1806 (von Platen, 1806), presented to Gustav IV. The king and the cabinet supported the idea of having the canal further investigated. Baltzar von Platen, who would become the primary entrepreneur in furthering and leading the following process, thus was given a royal assent in 1808 to go ahead with a more detailed investigation of the proposed canal, based on the previous investigations and the 1806 report. The reasons for the decision to go ahead with the canal project was probably different as seen by the pre-revolutionary regime and the post-1809 regime. Gustav IV had an interest in the general modernization of Sweden, where the canal project fit. The post-1809 regime had to grapple with a much more challenging situation with Sweden's geopolitical situation being changed following the loss of Finland in 1809–1810. In both these situations, Göta kanal could though play an important role. More on this below.

The investigation was carried out during the summer of 1808. A report was finally sent to the King early in March 1809 (von Platen & Telford, 1809), only three days before the dethronement of Gustav IV. A more detailed proposition with extensive maps and construction drawings was the main content of the report.

The report, naturally, was not attended to with the highest priority in the midst of the revolutionary process of 1809. As it turned out, von Platen though found himself to be closely involved in the processes of the revolution and, as a member of the nobility with close contacts to some of the leading revolutionary men, and also to the new king, Karl XIII, the brother of Gustav III, advanced quickly to become member of the new cabinet from June 1809. This was a position which offered good possibilities to influence the future policies of the country, but made it impossible for von Platen to act in the Riksdag.

The proposal from earlier 1809 was instead taken on board by some members of the Riksdag, who argued for a parliamentary decision urging the government to act in order to have the canal project confirmed and decided. More about this below. The outcome of the deliberations during the fall of 1809 was that the government proposed the building of Göta kanal in a bill of 10 November 1809. The process leading to the construction of the canal was set in motion during the spring of 1810, which will be further described below.

NOTE

1. Riksdaler Banco, the official currency in Sweden at the time.

REFERENCES

Bring, S. E., et al. (1922/1930). *Göta kanals historia, del 1–2.* Almqvist & Wiksell.

Carlsson, S. (1944). *Gustaf IV Adolfs fall: krisen i riksstyrelsen, konspirationerna och statsvälvningen (1807–1809).*

Hasselgren, B. (2018). *Transport infrastructure in time, scope and scale, an economic history and evolutionary perspective.* Palgrave Macmillan.

Von Platen, B. (1806). *Afhandling om Canaler genom Sverige med särskildt avseende å Wenerns sammanbindande med Östersjön.*

Von Platen, B., & Telford, T. (1809). *Berättelser samt kostnads-och ersättnings-förslag rörande den föreslagna Göta kanalen.*

CHAPTER 4

Different Perspectives on Göta kanal

A Long-Term Decision Process

This chapter is based on the multifaceted co-evolutionary model presented in Chapter 1, but goes further in elaborating different aspects, with a focus on themes related to primarily political, economic and organizational areas of influence.

The Göta kanal project had been discussed since the mid-1700s. A decision by Parliament to start the project was even taken by the mid-1700s though without any strong effort of implementation. When the final decision was taken in 1810 to actually establish the Göta kanal corporation and to start the construction works, the project was thus since long a theme of discussion, alongside a number of other canal projects. Earlier plans had fallen short due to lack of financing and other difficulties for the King and the Cabinet to set plans into realization, as discussed above.

One of the most obvious reasons for the renewed discussion on the canal project finally to take off in the early 1800s, but not earlier, was the situation that the connecting canal in Trollhättan had been finished some years earlier. It was by then more obvious that an inland waterway-connection between Lake Vänern through Lake Vättern to the Baltic Sea would have a potential of not only connecting the great lakes but also to bring a completely new transportation potential for the capital Stockholm at the east coast and the general trade flows between east and west in the country.

© The Author(s), under exclusive license to Springer Nature Switzerland AG 2023
B. Hasselgren, *An Institutional Approach to the Göta kanal*,
https://doi.org/10.1007/978-3-031-44416-6_4

Baltzar von Platen set the process in motion through publishing a report on the building of the canal in 1806.[1] The plans were based on earlier proposals but refined and renewed. Von Platen was not an engineer himself. He had though been involved in the construction of the Trollhätte kanal as one of the directors of Trollhätte kanal since 1798 and had undoubtedly good connections to the central administration and the government in Stockholm.

Von Platen was married to a daughter of one of the leading trading-house families in Gothenburg, Hedvig Elisabeth Ekman. The Ekman family would be instrumental in the later financing of the canal. And Hedvig Elisabeth's brother Gustaf Henrik was engaged as a member of the Board of Directors in the Göta kanal corporation. Ekman also was an important contact point for the connection to Telford.

One of the issues in the discussion on the canal and its geographical coverage and stretches was a long-standing dispute between the two cities at the eastern coastline; Norrköping and Söderköping. Norrköping was the major of the two, concerning trade, and argued intensively for the canal to have its beginning in the vicinity of Norrköping. Norrköping also pointed at the less favourable status in general of Söderköping as a place for future trade flows and growth. Söderköping on the other hand argued for its own priority over Norrköping. Lindgren (1993)[2] reports on these conflicts which, for a number of years, generated an aggravated debate and a number of petitions and letters to the government. This dispute to some part resembles debates in relation to debates in Britain in relation to canal and river projects reported by Bogart (2016). Interest groups furthering local interests indeed were organized in early nineteenth-century Sweden in a parallel manner to British experiences.

The first decision which led to the actual construction of the Göta kanal was taken by the King Gustav IV in 1808. Based on von Platen's early plans for the canal, an instruction was given to von Platen to continue with the plans. Based on this an engineer, Samuel Bagge, was involved in the further planning of the canal during 1808–1809. At this point von Platen also engaged Thomas Telford, the English leading canal engineer. The engagement of Telford was mentioned in the royal decision of 1808 according to Strömbäck (1993). Von Platen sent his introductory letter to Telford on 28 April 1808 referring to the royal decision on the canal project.

Samuel Bagge, the first director of engineering of Göta kanal, together with Telford, who visited Sweden during the late summer and early fall of

1808 on the invitation of von Platen, were instrumental in setting up the next and more detailed plan for the construction of Göta kanal. Telford investigated the suggested canal stretch and proposed some alterations to the plans. A new plan for the construction, with additional locks and some other changes, was produced during 1809.[3] The cost of the project had more or less doubled compared to von Platen's 1806 estimates, from ca. 830.000 Rdr to ca. 1.600.000 Rdr.

A Strong Leading Proponent

Göta kanal is clearly an example of a project that would not have been realized without the continuous and strong leadership and vision-generating abilities of the main initiator and later leading director and chairman of the Göta kanal board Baltzar von Platen. With his profound and lasting contacts in the government and serving as a cabinet member for a couple of years following the 1809 revolution, his access to the King and the cabinet was unusually well developed.

Von Platen also managed to connect the project to important actors in Britain, not least to Thomas Telford,[4] when it came to acquiring the necessary planning and construction skills for the realization of canals, and he also managed to build alliances with centrally placed actors in Britain involved in canal construction over time.

It seems like von Platen had a strong ability to communicate the messages and visions of how the canal would work as a tool for restoring the national pride of Sweden but also something that would increase the wealth by its trade and growth-enabling effects. In this way, von Platen resembles a Schumpeterian entrepreneurial figure who gathers the necessary resources; financial, technological as well as organizational, to make a project viable. Von Platen also had a strong role as a "political entrepreneur". In this role he managed to raise support in the necessary situations in the government/cabinet and in the Parliament, whenever the project was under scrutiny or the aim for political attack from a successively livelier opposition during the canal project's lifetime. The continuous success in bringing together the necessary resources and support was at the centre of the entrepreneurial skill that von Platen showed.

At the same time, it should be remembered that von Platen, though being a successful political entrepreneur when it came to vision formation and establishment of support and coalition-building had less obvious organizational skills himself. The organization of the corporation's internal operations show signs of repeated situations when the necessary clarity and structure of decision-making seems to have been lacking. The furthering of the long-term vision seems often to have been put in the first room by von Platen, while the everyday work of managing and controlling the practical operations might have been less important to him, or executed with a basis in a strict hierarchical tradition acquired in his military training. Perhaps, some of the financial crises and substantial delays in the canal project could have been avoided had the organizational efficiency and structure been more in focus of the Board of Directors of the Göta kanal corporation.

The project might very well have been realized also without von Platen. Examples from other countries show that different organizational and financial structures were used. It is though reasonable to believe that without a strong proponent as von Platen, the initiation of the process might have been delayed. A delay for some decades would have made the canal plans to come closer to the introduction of railways, a discussion that was initiated in the 1840s in Sweden. This would probably have made it difficult to raise sufficient support for this major project.

The Constitutional Framework for Building a Canal

It might be seen as a sign of the importance of the political situation that surrounded the Göta kanal project that the first lock of the first opened "Western line" (from lake Vänern to lake Viken in the east) of Göta kanal seen from Lake Vänern originally was named *The Constitution*, with the following four locks originally bearing the name of the four estates of the Swedish Parliament (Riksdag) at the time.[5] On the other end of the western stretch of Göta kanal, the first lock, at Forsvik, is named after the King elected following the 1809 revolution, Karl XIII.

This first stretch of the canal is thus an illustration in itself of the political and constitutional framework established by the Swedish 1809 revolution. Göta kanal has a history closely connected to the political economy environment of the time, even if there is of course nothing directly connecting the two just because of the naming of the locks (Fig. 4.1).

Until 1809 the constitutional framework (The Instrument of Government) was established, and from the revolution in 1772 led by King

Fig. 4.1 The first lock on the Göta kanal, from Lake Vänern to the east, originally named "The Constitution" (*Photo* Björn Hasselgren)

Gustav III, the King enjoyed a wide area of independent decision-making powers according to the constitution of 1772. With the adjoining Act of Security and Union established in 1789, a rule of royal supremacy became a fact. The balance of powers was arranged in a way that gave the King a fairly wide area of decision-making powers, without having to appoint a Royal Council, which Gustav III decided not to do, and a more limited role of the Parliament compared to the former "Age of Liberty", from 1719 to 1772.

Following the dethronement of the King Gustav IV in March 1809, a revised Constitution Act (1809 Instrument of Government) was established in June 1809, with a more balanced division of powers between the main three institutions: the King, the Parliament and the Courts.

A main difference in practice between the two constitutional settings was perhaps the number of times the Parliament was summoned. From 1792 until 1809, the Parliament thus only met on one occasion, in 1800. The unwillingness of Gustav IV to summon a Parliamentary session in 1808–1809 to resolve the difficulties following the Finnish war, was one of the main reasons for the Revolution of 1809, though among many others.

In the pre-1809 setting, the King and his powers were in the centre of the institutional setting. Following the 1809 revolution, the initiative was changed in the way that the King still had the role to be the active ruler of the Kingdom, while the Parliament had the right to decide on new legislation and taxes. The 1809 constitution also made it compulsory for the King to establish a Cabinet to council him in his decisions. The Parliament also regained the powers from pre-1772 to execute reviewing and auditing roles in relation to the King and the Cabinet, besides deciding on taxes.

The establishment of a large project like the Göta kanal was a question, both pre and post 1809, which would require royal assent in some way and following 1809 also a Parliamentary approval as regards financing and the need for specific legislation.

It is not obvious whether von Platen at this stage had a plan for the future organization and financing of the project. In the 1806 report, von Platen included a rough estimated cost and income calculation, that indicated a possible future profitability of the canal, based on the (very rough) cost estimation of ca. 830.000 Rdr,[6] and with a coordinated income structure, where receipts from Trollhätte kanal, already in operation, should be shared with Göta kanal.

A discussion on the possible organization of the canal project as a business corporation, with shares issued widely in the country is discussed, with inspiration from, for example, the Trollhätte kanal operations. Von Platen discusses whether the calculated return on the capital invested would be enough to attract money investors, but concludes that the estimated return on investment, 1.80%, "is a dividend which would not attract any Capitalist, who only seeks return on his capital".[7] Other sources of income would therefore probably be necessary, why von Platen foresaw either lending from the state on favourable conditions, or as a second-best alternative, borrowing from abroad, in specific if a lottery would be the basis for dividends or interest payments.

Von Platen, according to Bring (1922), actively sought any possible way of furthering his proposal's realization through different contacts in the central government administration. By late 1807, he was contacted by the King's state secretary, Mattias Rosenblad, who asked von Platen to meet with the King. The meeting was held early in 1808 and a formal decision by the King to support a further investigation of the canal stretch and the entire project was taken on 12 February 1808. This is an example of the direct involvement of the King in a wide array of matters, as a result of the division of powers in the pre-1809 constitutional environment.

Von Platen was indeed well aware of both the constitutional and regulatory environment in Sweden and the canal-building experiences from Britain and other countries. Looking at the contemporary examples of other canal projects, primarily in Britain, the common way of organizing canal enterprises was as limited stock corporations. Trew (2010) points to the most usual process for organizing a canal project in England, under the 1720 Bubble Act[8]: a bill had to be passed in Parliament. From 1794, further specifications were necessary for such a bill to be passed according to Trew: a map of landholdings, reference books of landowners and occupiers in the vicinity of the canal with a record of their support or opposition to the proposal and, finally a list of financial supporters.

The 1809 report[9] entails many of the aspects that were stipulated in the 1794 specification of the Bubble Act in Britain. Descriptions of the stretch more in detail compared to the 1806 report with maps and renewed cost estimations were thus included. This time the cost estimation ended at a total cost of 1.597.491 Rdr, nearly twice the original cost estimation from 1806. Questions relating to the organization of the canal project were though not among the ones that were prioritized during the investigation of the canal stretch in 1808–1809. Von Platen and Thomas Telford travelled along the planned canal stretch during the summer of 1808 and Telford's annexed report to the 1809 report is dated 25 September 1808, at Frugården. Frugården was von Platen's private home and residence since the late 1790s and is situated at Lake Vänern, some 400 km west of Stockholm.

The final report is dated 10 March 1809 by von Platen and formally written as a report to Gustav IV, who was dethroned in Stockholm only three days later. According to Bring (1922) the report was printed during the spring of 1809 and handed over to the members of parliament in Stockholm in May/June 1809, by then assembled as a constitutional assembly under the new regime. The adjoining maps of the proposal's stretches were printed in 1810 and give detailed descriptions of the canal and its proposed locks.

The canal project was carried over from the earlier Gustavian regime to the new constitutional setting with a surprising easiness. It would be easy to understand if the new regime, in the light of the generally poor state of the state finances following the Finnish war, would have been inclined to either cancel or at least pause the plans for the canal. Instead, it seems like the new regime sought for a project like the Göta kanal to divert interests to, perhaps as a new national project for the country to unite around. Therefore the 1809 revolution seems not to be a crucial dividing line for the project as regards its continuation, as it could have been expected to become, in the midst of a drastic political and national upheaval.

The fact that von Platen managed to further the project both during the old and the new regime is something that has been taken note of in the traditional historic sources discussing the canal. It has not, though, been a source of any extensive elaboration. The research in relation to the last years of Gustav IV's regime, the revolution and the elaboration and installation of the new restored regime is vast. Carlsson (1944) and Björklund (1965) give important insights into the events, where the details in the process leading to the constitutional change are focused by Carlsson, while Björklund is more centered around the different proposals for the new constitution circulating before and after the dethronement of Gustav IV.

Von Platen seems to have been one of the figures mastering the political changes with excellence. Clearly, he was well connected on both sides in the events. On the one side, he is stated (Bring) to have been a supporter of the Gustavian regime. On the other side, von Platen had early and close contacts to one of the central figures in the revolutionary process, Georg Adlersparre. Adlersparre was among the few people that von Platen sent his 1806 Göta kanal proposal to for comments, even though Adlersparre at that time had no formal role in the administration, at the time being one among the opposition groups arguing against the regime in various ways.

Adlersparre also visited von Platen in 1808 at his estate *Frugården*, soon before Thomas Telford arrived to take part in the 1808–1809 investigation of the canal. Adlersparre is referred to having brought to the fore the need for dethroning Gustav IV in order to bring about an end to the peoples' suffering. Von Platen is attributed with having opposed to this proposal (Carlsson 1944). When it came to another strategic question, on the possible union of Sweden and Norway, however, von Platen and Adlersparre seem to have shared similar views.

During the decisive days in early March 1809, when the revolution was carried out in Stockholm and Georg Adlersparre was leading the part of the Western Swedish army, he was in command of, towards Stockholm in order to press for a change in government, it though seems like von Platen was entirely inactive when it comes to the political upheaval. It could have been that the news of the political change did not reach Frugården, where von Platen lived, until later and if the dating of the 1809 report is to be followed presumably, von Platen was probably engaged in hectic work finalizing the report.

According to Carlsson (1944), the support of the revolution outside Stockholm was also quite weak. Pro-royal views seem to have been kept for some time by the heads of the government county administrations and bishops in the major regional towns. Carlsson argues that Linköping was one of the towns where the revolutionary support was weak or even anti-revolutionary views the stronger. Gothenburg is another town, where support for the revolutionaries was lacking. It can be noted that Linköping and Gothenburg were to become two of the towns with the strongest support for the Göta kanal project in the coming time. Since 1800, Baltzar von Platen was married to the daughter of Gothenburg's most wealthy trading house owners, Hedvig Elisabeth Ekman. Von Platen might have had good reasons to be working on his report on the very last days of the old regime.

The news of the revolution though clearly had reached von Platen by 28 March 1809 as he writes to Telford from Frugården to update Telford on the latest events ensuring Telford that the "change in government seems as yet to occasion very limited disturbances in the country".[10] Could it be that the political situation had become clearer over the last few weeks, making a statement on the future of the canal project possible? And perhaps also to support the revolution and what it would bring seemed to be a safer strategy than earlier in March? Von Platen over time had the ability to position himself with the winning team.

Later on, during the 1809 events, it is interesting to see how von Platen, being a member of the Nobility estate in the Parliament, managed to be involved in the inner circles of the small group of men often to be named "the men of 1809", shaping the new political scene and constitutional order. Bring (1922/1930) points to the fact that von Platen became engaged by the new regime as early as in April 1809 for a number of delicate tasks.

One of them to try to underscore the old rumours from earlier decades that Gustav IV was not the legitimate son of Gustav III, a project which though more or less failed. The second to ensure the continued support of the British navy operating west of Gothenburg, which was more of a success, since the British navy continued to circle around in the vicinity of Sweden for the time to come. The third to try to confirm the interest of the Danish Prince Karl August, Danish Governor to Norway, for the role as heir to the Swedish throne. Also, this seems to have been a success. All these three assignments were carried out by von Platen from March until the end of April. This clearly is a sign of von Platen's willingness and eagerness to take part in the matters of the new government to be formed. It seems unlikely that he had not, even if silently, been supportive to the views held by Georg Adlersparre and others that a political upheaval was necessary for Sweden's future development.

Not only is von Platen already before the revolution an influential member of Parliament, he was also elected to become a member of the Constitutional Committee with the role of proposing a new constitution, during a few hectic weeks in the spring of 1809. Von Platen was also elected to the first Cabinet according to the new constitution as of 9 June 1809. Von Platen was to remain in this position until 1812, when he left the Cabinet following a dispute on the tolls on domestic trade to be reintroduced.

During the 1809–1810 parliamentary session, several members of Parliament were in favour of the government to act more decisively to propose to the Parliament the canal project be voted on and initiated. Among these are representatives of Linköping, Vadstena and Gothenburg all three cities that would be favoured by the project. Here some of the influential actors in the coming canal debates and in the corporation took important roles, such as the wealthy trading house owner Berndt Harder Santesson in Gothenburg, Bishop Carl von Rosenstein in Linköping and Mayor Erik Gezelius in Vadstena. Bring (1922/1930) points to the close relationship between von Rosenstein and von Platen since their experiences in the navy in the 1780s. Santesson was connected to the Ekman family, and Vadstena was the town closest to the canal's planned connection to Lake Vättern on the eastern stretch towards the Baltic Sea.

The four estates of the Swedish Parliament (nobility, priests, burghers and peasants) finally were united in a demand to the government to come back to the Parliament with a proposal for the realization of the canal. A bill was eventually presented in November 1809. It seems likely that the bill was written by Hans Järta, at the time finance minister and von Platen, two of the leading "men of 1809" and both participants in the Constitutional Committee and in the Cabinet. Järta and von Platen also shared the view that continued and close contacts with Britain was a central interest for Sweden, no matter what was stipulated by France, which was the favoured international partner for most influential figures in the administration in Swedes at the time.

Following continued discussions in the Parliament and the Riksens ständers bank (the Central Bank) primarily on the conditions for the state lending to the Göta kanal corporation planned to be established and on the regulation of the financing arm of the Göta kanal corporation, the Göta kanal Diskont a final decision by all estates was taken on 22 February 1810 to initiate the project. The Cabinet had already decided to set an organization committee for the formation of the corporation in operation as early as January 31, and by the end of February the process was thus finally set in motion. The committee had as one of its main tasks to arrange the issuance of shares in the new Göta kanal corporation. More on this below when discussing the financing issues in relation to the canal.

On a number of occasions, the project would be contested and criticized by the Parliament in the years to come. The King or the parliamentary majority though, in the end, always came to the managing director and chairman of the board von Platen's and the project's support. With this wider background, it seems quite justified that the locks of the first stretch of the canal were named after the Constitution and the Estates, as well as the King.

The canal had become something of "a national corporation", just as von Platen fairly soon would argue that it was. The 1809 revolution was important for the project as regards the arguments used to support it and the way the project was carried out. Without the revolution, von Platen would still probably have had important support for the project with the former King and senior officials and trading interests backing the project. The influential roles that von Platen was able to take following the unexpected sequence of events in the spring of 1809 though put him in a better role to realize the plans.

FINANCING OF THE PROJECT

The project was financed through a combination of private sector financing and substantial support from the state. Over time the balance between these funding sources varied due to political and project specific reasons. The Göta kanal Diskont was established to support the Göta kanal corporation from the outset, and was important as a financial vehicle for the realization of the project, though dismantled in 1817 following a banking crisis in a sister bank external to the canal project. Up until then, the Diskont was a formally separate but partly integrated operation with the canal corporation. Shares in the canal corporation could be used as security for borrowing from the Diskont. Share subscription could thus be seen as a form of deposits in the Diskont.

The organization of canal projects in corporate form had been tried earlier in Sweden. Both the late eighteenth-century projects Strömsholms kanal and the Trollhätte kanal were set up in this way, with a combination of private and state sector financing. The difference compared to primarily the Strömsholms kanal was that that project more closely confined to the interests of merchants and industrialists who had direct business interests in the canal to be constructed. These actors took the initiative to the project and also held the majority of the shares in the company, backed by government support.

Government support to the Göta kanal project was channeled through a mix of direct government funds, guarantees from the government and successive support decisions transferring additional financial support to the corporation in different ways. The other major source of financing had its origin in shareholders' paying for shares in the company. Compared to the earlier canal projects, the issuance of shares was much wider based on a public offering to sign for shares and the final holding of shares was spread to all different regions in Sweden, and among a wide set of actors in different social strata. The Göta kanal shares might be seen as one of the first modern widely held stocks in Sweden. All through the construction process, the project was organized as a formally private corporation with continuous shareholders' annual meetings and an administration that was formally separated from the government.

Important in any major construction project in the field of transport infrastructure is the handling of different risks. Planning and design risks in the case of Göta kanal were (more or less) taken by the corporation and subsequently led to a substantial need for government support as the risks materialized.

One of the major risks in relation to construction is the possibility to acquire land for the construction. This was regulated in the government decision of 1810 where the Göta kanal corporation was given the right to control a specified land area on both sides of the planned canal stretch and even to fence it in. Special regulations were also established for the valuation of land that the corporation had to acquire, set out in a specific government decision.

The Göta kanal Diskont was set up as an integral organizational unit of the Göta kanal corporation, but with a separate Board of Directors, though with central actors participating in both Boards. The Bank could use a Central Bank (Riksens ständers bank) guarantee as a general support for its operation. It received deposits from the public and extended loans to the public. There was, at this time, in general a liquidity surplus in Stockholm and a borrowing demand in Gothenburg and other parts of the country, for example in the regions surrounding the future canal. According to Bring (1922/1930), ca. 70% of the lending of the Diskont in 1811 was extended to borrowers in the adjacent regions to the canal works. Borrowing from merchants/trade and farming were the two dominating sectors in the lending stock at the same time.

Lack of banknotes was seen as a general problem in the early nineteenth-century Swedish economy. The Diskont, together with the two other Diskonts at the time, in Malmoe and Gothenburg, were instrumental in supplying notes to the economy. The Diskonts though did not operate according to any clear-cut rules as to the proportion of notes to be extended in relation to deposits and the government guarantee in the Central Bank. Thus, deposit notes were used as notes in the public domain as was also the borrowing certificates that were extended by the Diskont as currency for loans in the Diskont. In addition to this, notes were also extended in relation to the Central Bank guarantee, which (all in all) summed up to amounts well above the levels of own capital or government guarantees.

In a market with surplus of liquidity in some geographical areas and general inflationary pressures during the Napoleonic wars, Bring argues that the offer to participate in the formation of the Göta kanal corporation through signing for shares was attractive. The issue of the shares was also, as it turned out, very popular with demand for three times the issued amount of share-capital. One of the strong arguments in support of the offer was of course the goal of the Diskont to pay a yearly dividend of five percent on the shares in the Göta kanal corporation, also mentioned in the 1810 royal charter of the Göta kanal corporation, thus giving it a kind of government guarantee.

Even if the Diskont made a good contribution to the financing of the canal costs soared well above the available financing. Another problem was that the King, from 1818 Karl XIV Johan, was personally deeply interested in financial matters. Carl David Skogman, the finance minister to become, who had travelled to Britain and the US to study financial affairs and trade on behalf of the government, and who returned inspired by the liberal market economies in these countries, was the Diskont's director in Stockholm. Skogman had a role to mediate between the King and the Diskont. As the Malmoe Diskont, which was most heavily expanding its lending business into financing of more risky trade-related operations, was (more or less) bankrupt in 1817, the two other Diskonts were also brought into financial distress.

The King is reported[11] to have disliked the Diskonts since he believed that the more or less unregulated issuance of bank-note equivalents would be harmful to the economy, inducing increased imports and supporting speculation against the Swedish currency. The difficulties in the Malmoe Diskont were a reason for the King to act in order to dismantle all of the three Diskonts. This was effectuated by the lack of "lender of last resort"-support by the government in the fall of 1817, and the theme for debate and decisions in the extra Parliamentary meeting in the fall of 1817.

The three Diskonts were dismantled following the decisions. The operations were taken over by Riksgäldskontoret (the National Debt Office) which served as bank resolution institution. The dissolvement process for the Göta kanal Diskont lasted for the following 30 years. It was indeed a soft landing, and also a sign of the widespread operations of the Diskont, following the 1810 widespread issuance of shares and subsequent voluminous lending activities.

For the Göta kanal corporation, the decision to dismantle the Diskont was near to an existential crisis. The Diskont was mentioned as an "inseparable" part of the canal operations at large in the establishment decree,[12] and the Board of Directors sent a formal harsh protest to the government following the unwinding decision. The protest was, however, not reacted upon by the government. Following this process, the Göta kanal corporation, though formally still a separate entity in relation to the State more and more became dependent on financial support from the government. Subsequent loans and support in other forms were also given to the corporation.

In the end, as the kanal was finished in 1832, the total private sector financing of the project was substantial. 3.141.200 Rdr (ca 30%) of a total cost of 10.385.799 Rdr[13] had been paid by the shareholders. Göta kanal was indeed an example of private–public financing collaboration.

On the Issuance of Shares, etc.

Already in the 1806 report, von Platen (von Platen 1806) outlined a plan for how to attract capital for the project by a subscription process preferably to be organized in many places in Sweden, primarily those that would have a direct interest in the project. The organization of the subscription should, according to von Platen, be given to locally trusted men, who should be able to exchange information continuously on the outcome of the subscription process.

This projected process was also more or less followed as the Göta kanal was discussed in Parliament during 1809–1810 and the canal-corporation eventually established in the spring of 1810. Something similar to an investor memorandum was put together by the organization committee in which the proposal and investigations supporting the decision by the government to issue the authorization of the corporation were included. This material was communicated in the press in April 1810. Subscription offices were organized in a number of towns all over Sweden. The subscription period was kept short but the offered share capital was over-subscribed as mentioned above.

The data in Fig. 4.2 shows that the shares were primarily subscribed in Stockholm, Gothenburg and the town of Vadstena, with the major royal navy town Karlskrona as the fourth largest town of subscribed amount. For Vadstena, the subscription was a very large amount compared to the financial resources of the town's inhabitants. The picture of a project, which is dominated by Stockholm and Gothenburg interests, but with widespread support around the country is apparent. It should be noted that only a small proportion of the subscribed amount had to be paid at the time of subscription, some 10–15%.

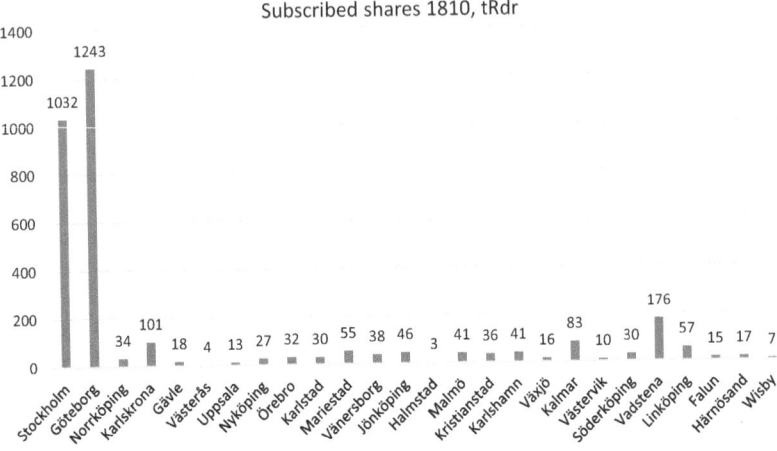

Fig. 4.2 Subscription for shares in the Göta kanal corporation April 1810, places where subscription lists were presented, tRdr (*Source* Göta kanal archive, Vadstena, GIII c1 1–4)

The shares in the corporation could, as reflected above, be used a security for borrowing in the Göta kanal Diskont. No data has been found on different securities delivered to the Diskont as security for borrowing in the Diskont and the data available on lending is not exactly comparable to the towns presented in Fig. 4.2. Data for lending to counties instead of to separate towns have been used here and split evenly between the towns in the same county in Figs. 4.2 and 4.3 in order to give a picture of lending to the company and borrowing from the Diskont.

Figure 4.3 gives an indication that the borrowing need was different from the willingness to subscribe for shares. Large lending from the Diskont took place to borrowers in Göteborg (Gothenburg) and the towns in the vicinity of the canal to be built. Stockholm though shows low amounts of borrowing from the Diskont. A hypothesis is that liquidity surpluses in Stockholm were transferred through the Diskont, and through the corporation, to borrowers in other geographies adjacent to the canal. Borrowing in the towns surrounding the canal might also be a sign of increased business activity in these regions/counties.

In Fig. 4.4, finally, the subscription of shares in the Göta kanal corporation (here primarily to be seen as an obligation to finance the corporation

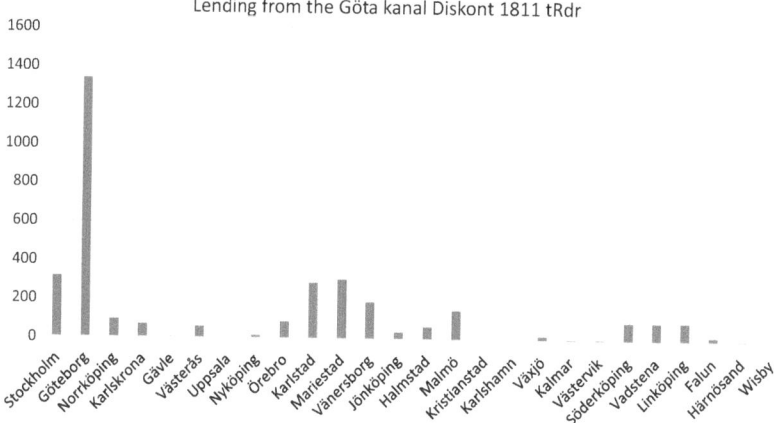

Fig. 4.3 Lending from the Göta kanal Diskont in 1811 tRdr to borrowers in different towns (counties) (*Source* Bring, 1930, p. 148. Some data compilation has been made to match data for counties with towns)

in the future as the subscribed value of shares had to be paid) is compared to the borrowing from the Göta kanal Diskont in different towns. This presentation of the data gives a slightly different pattern. Stockholm still is the main source of net financing. The major "net borrowing" towns (or more precisely individuals residing in these towns or in the surrounding region) though seem to be spread over the country with towns around Lake Vänern as the most borrowing, but also with towns not directly geographically connected to the canal as net borrowers. Inhabitants in Vadstena, which was a large supplier of share-capital subscriptions relative to its size, was not borrowing to the same extent as other small towns. Perhaps, this is a sign of another type of capital, i.e., long-term savings, being invested in Vadstena compared to the more trade-oriented towns in western Sweden. This would though have to be further investigated.

Actors representing financial interests with different geographical locations were connected through the financial intermediation between lenders and borrowers and regions through the Göta kanal corporation and the Diskont. In the years to come, these relations would play out in different ways and would also blend with the interest represented in the government and Parliament in varying ways over time. More on this below.

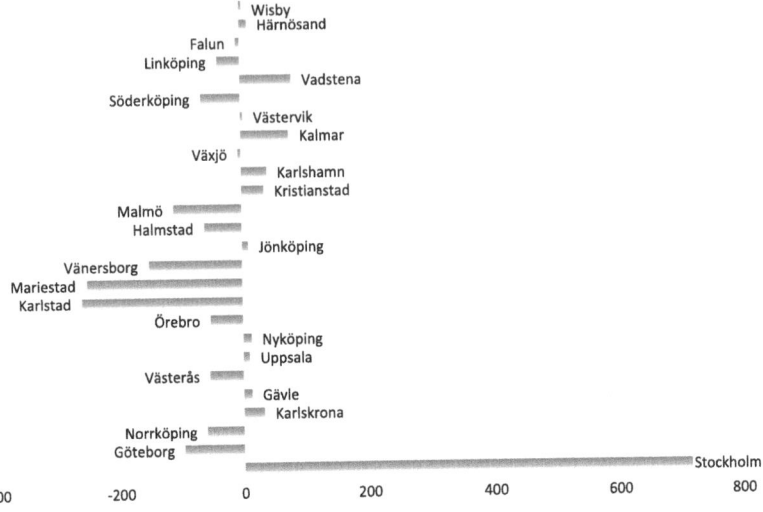

Fig. 4.4 Subscriptions 1810 less lending from the Diskont 1811, tRdr (*Sources* As in Figs. 4.2 and 4.3. Own data compilation. Some data compilation has been made to match data for counties with towns)

ORGANIZATION OF THE CANAL PROJECT—INDEPENDENT OR "A NATIONAL CORPORATION"

The Göta kanal corporation was formally an entity independent from the government. It also acted independently from the government in its day-to-day business, even if the close personal ties between the Cabinet, the central government agencies and the Board of Directors were obvious. The guiding principle from the start to form an independent organization was, however at the same time, kept all through the project. One of the reasons for this organizational structure has been discussed by Gunnar Petri (2017) in his biography of one of the important actors in the early days of post-1809 government, Hans Järta.

Järta had been secretary in the Parliamentary Committee developing and formulating the new constitution act during the spring of 1809. Järta was inspired by liberal ideas furthered by Adam Smith and other scholars in the Scottish Enlightenment tradition. Also having a long experience from the central government in Sweden and now in the position as the equivalent to finance minister, Järta was the man who drafted the 1810 decisions on the establishment of the Göta kanal-corporation.

According to Petri,[14] Järta was convinced that the incentives for efficiency in a project such as the Göta kanal would be stronger if it was established as a formally private corporation with anyhow substantial private financing compared to as a government agency financed with government borrowing and appropriations. Studying some of the difficulties the corporation experienced over time, it is also clear that the leading figures of the corporation, apart from being strong proponents of the project, also felt a strong personal responsibility for the continuation and completion of the project. Järta might have been right in his analysis of the strength of personal responsibility as an incentive for efficiency. But it is also relevant to add that there were experiences from the earlier canal projects to refer to, which had been carried out as corporations with government support.

It can though of course be questioned whether there was in fact any independence of importance for the corporation in reality. On the one hand, one can argue that any major business corporation at the time would probably have had to be based on some kind of royal or government consent or even "privilege" as in the case of Göta kanal. This would speak in favour of seeing the corporation as just as much independent as other similar ventures at the time, though of course few were at hand. The fact that a substantial amount of capital was paid, or at least promised to be paid at a later stage by the shareholders, speaks in the same direction. Göta kanal was a relatively independent business corporation.

On the other hand, the close connections between the government/cabinet and the corporation and the fact that the Diskont, which in turn financed the corporation, enjoyed a special Central Bank guarantee, points in another direction. There were both obvious and silent obligations on behalf of the government to give its active support to the canal project, by extending guarantees, offering (primarily military) staff and through a generally positive view on the entire project. The connections between the government and the corporation would though be challenged during a number of crises and difficult situations during the construction phase.

A question that would be reiterated over time during the construction of the canal was the actual formal status the project had and thus the degree of support it should count on or be eligible to receive from the government. An indication that the government had a wider perspective on the canal corporation than just as a business operation is indicated in the royal decision of the establishment of the corporation in April 1810. The decision points to the positive "external effects" the canal would hopefully create, in addition to the narrower financial perspective of the operations of the canal as business in itself. The canal was thus expected to further agriculture and support other sectors of business, bring a livelier trade and business climate in general, and to foster industriousness at large. This theme will be further discussed in the following chapter, where the relation between the canal works and the adjacent iron industry is investigated.

The canal was also expected to form a main structure in a future possible system of feeding canals in the regions connected to the canal stretch. This "socio-economic" view on the canal was also the theme of various speeches held by the Director von Platen in Parliament (as a member of the Estate of Nobility of the Parliament) and in a number of letters to the government and the King.[15] Von Platen had the aim of turning the Göta kanal project into something of a "National business operation" ("Nationalföretag"). Over time, the situation also more or less developed into such a privileged status of the corporation.

In some way this was a contradictory development, since the initial intention of von Platen was likely to raise political and financial support for the vision of the canal project, which should though in practice be operated as an independent body. As the challenges to the project mounted over time, however, the need for government support to finalize the project became apparent, and the close ties to the government became more a necessity than a general advantage. This can be seen as a parallel to the situation of the state railway corporation in Sweden during the twentieth century where the increasing financial distress of the railways over the century was combined with claims from the railway corporation that the operation was of general public interest and importance.

As regards the organization of the Göta kanal corporation, it is interesting to see that the Diskont and the corporation were established as "inseparable" entities. They had overlapping Boards of Directors and the Diskont can in many respects be seen as the financing department of the corporation. The Diskont was responsible for the cashier's function of the operations and attracted deposits both from private depositors and from government agencies in order to keep a sufficient liquidity in the entire operation. The possibility for shareholders to borrow in the Diskont with the shares of the corporation as security for the loan also displayed the strong bonds between the two arms of the combined canal corporation.

It seems reasonable to see the Göta kanal corporation as part of the government's area of influence, and to a growing extent, during the construction phase from 1810 until 1832. Over time the corporate structure was though kept. It seems like the finance minister at the time of the establishment, Hans Järta's alleged view that a separate corporation would strengthen the responsibility of the corporation, was supported over time. Of course, attempts to alter the organizational structure would also have put the government in a situation where the shareholders would have had to be compensated in some way. Without doubt it would have generated tremendous difficulties for the government to raise the necessary funds for such a bail-out of the shareholders.

The (at least theoretical) option for the government to declare the corporation insolvent and to take over the corporation without any compensation for the shareholders, was also probably a less attractive option, which may have caused a major financial distress among the influential shareholders. This probably led to the "middle way" the government choose of both engaging in extended financial support to the corporation and to preserve the formal private ownership of the corporation. The extended winding-up period of the Diskont, around 30 years, is also an example of the cautious way the government treated the financiers and borrowers of the corporation.

THE CONSTRUCTION PHASE

The Göta kanal corporation immediately was set into operation following the formal establishment. Construction works commenced during the summer of 1810 in a number of places along the canal stretch. The organization was for administrative and practical reasons divided in two organizational units, west and east. Lake Vättern was the dividing point between the two organizational units. This way of organizing the project resembled the Caledonian Canal project.

The construction phase was planned to take eight to ten years, according to different estimates presented during the preparatory phase. In the decision of 1808 to go further ahead with the project, the King at the time had though pointed out a construction time of six years. However, it soon became obvious that there were substantial additional difficulties connected to the construction work that had not been realized in the planning of the project.

Reasons for the delays can be found related to many different areas and activities of the project. One thing that was marked by Bring[16] was the lacking number of available military staff as workforces compared to the estimated need, though these resources had originally been promised the Göta kanal corporation in the 1810 decision by the government. Thus, during the first years of operation, the military staff counted to 1.000–4.000 during the operation season from April/May until October. Following a request from the Board of Directors of the corporation, the military workforce from 1812 was though increased to around 7.000 men and a similar number of military staff arrived for the 1813 season, however only to be withdrawn following Sweden's entry into the Napoleonic wars. A number of military staffs of around 7.000 seem to have been more in line with the expectations of the Board of Directors than the more limited numbers of the earlier years.

The Göta kanal corporation thus faced a situation where staff had to be hired on the market for the continuation of the construction work. According to Bring, this was solved by hiring workers from the northern Swedish region of Dalarna which traditionally has been known for internal "export" of workforces to different purposes in Sweden. The staff numbering 2.300 in 1813 and 1.560 in 1814, included a rough hundred military staff and, according to Bring, an additional similar amount of Russian military deserters.[17] Even if the corporation had to pay for living expenses also for the military staff, the cost of hiring workforce on the market was far higher. Bring reports that the cost was above the estimated for 1813 and 1814 and on top of that food acquired for the far larger staff of military staff had to be re-sold with considerable loss.

Von Platen, summing up some of the experiences of the canal project from the first twelve years in 1822,[18] also argued that the use of military staff was essential for the project, from a pure project management perspective. The huge number of, largely untrained, staff necessary for the execution of the excavations, etc. was, according to von Platen, very difficult to organize in any other way than according to military organization and discipline. This entailed, von Platen argued, both the effective use of staff and other aspects like housing, sanitary aspects and food supply. Here, von Platen's earlier experiences as an officer are evident, but also the lack of civil construction knowledge and management in Sweden at the time.

In the midst of the challenging situation in 1813, Thomas Telford was once again sent for in order to evaluate the situation in the project.

According to Strömbäck (1993), Telford visited the construction sites for more than a month in the summer of 1813. The report by Telford was positive as regards the general activities in the project. Telford though recommended that additional skilled workers should be hired by the corporation. This was also one of the results of the visit by Telford. An extended period of hiring of English engineers and workers with experience from canal construction was initiated. According to Strömbäck, who has constructed a time series over the number of qualified English workforce in the project, primarily foremen, between 4 and 13 of these were constantly engaged from 1813 until the finalization of the project in 1832. At the same time, it is of course an open question to what extent a few managers could balance the lacking skills among a huge number of workers. Perhaps the effect of knowledge import from Britain is exaggerated by Strömbäck.

Costs had also soared during the project and the corporation found itself in a situation of financial distress late in 1814. The coming year a debate was held in the Parliamentary session (1815), following a special audit and a report from a parliamentary committee on the project. This marks the beginning of a period of criticism towards the project from many members of Parliament. Following a harsh debate concerning the government's proposal to extend direct subsidies to the Göta kanal corporation, the final decision was that the corporation would receive a yearly government loan of 300.000 Rdr for the following estimated ten years of construction to be amortized upon the opening of the canal.

The immediate financial crisis for the construction project was thus handled. For the continuation of the project, and there would be a number of new debated and crisis-alleviating decisions to come, the opposition in the Parliament was set to continuously criticize the project, both in Parliament and in the public press. It is probably fair to see this as partly a general critique of the project and government spending in general and partly as an object to attach government opposition to.

The management of the canal project seems to have been carried out in a balance between fairly modern project management practices contrasted against strategy and tactics according to the political situation at the time. Yearly workplans were made up for the coming operation and sent to the government. The workplans were fairly detailed and comprehensive. They included many different activities that had to be coordinated. It is difficult to assess to what extent the plans were actually followed. On several occasions delays and difficulties of different origin made cost and time estimates to become out of date.

Following the death of the chief engineer of the canal corporation Samuel Bagge in 1814 a new chief engineer, Erik Hagström, was hired. Hagström stayed in his post until 1820 when Gustaf Lagerheim, one of the assistants of Hagström was hired as new chief engineer. A number of revised plans for the project were formulated by Hagström and Lagerheim, while costs continued to increase and time frames for completion were adjusted. Substantial parts of the canal were though ready in 1822 (the western section) and 1825 (substantial parts of the eastern section). 1828 was though another year of financial crisis. The Parliamentary session of 1828–1830 finally decided to extend another 845.000 Rdr in support on condition that the project should be finalized in 1832 and that the shareholders should deposit 33 % of the amount with the National Debt Office as security. This final plan for the completion and financing was kept. By the completion of the project, though Baltzar von Platen was dead, he died in 1829.

The Role of the Göta kanal Project in Relation to National-Identity Shaping

The role of the Göta kanal project as part of the reorientation of Sweden's interests and geographical focus following the loss of Finland in 1809–1810 has been discussed above. Here this discussion is continued with the focus on the national-identity concept in focus. There is a wide literature on different core concepts and many different views on how national identity, if at all a viable concept, can and might be interpreted and shaped. Wodak et al. (2009) offer a thorough exploration of the discursive aspects of the term "national identity". Whether there is a link between transport infrastructure and the perception of a national identity is another crucial aspect to consider. National identity might of course be a concept that is separate from transportation-related aspects.

At the same time, and this is the view taken here, it seems more or less obvious that transport infrastructure has an important role to play in relation to the construction of and impression of a geographical area such as a nation. Major changes to the infrastructure system alters the way the nation is perceived and the way it is "used" by its inhabitants and others. In relation to transport planning, this is discussed by relating additions to the transport infrastructure system to the increased availability (e.g., shortened travel/transport time) the addition might lead to. Shortened travel times relate to a change, and hopefully an increase, in the economic geography of the area being analysed.

The construction of canals in the eighteenth–nineteenth centuries was indeed such a means of altering the structure and perception of nations, often in parallel with nations being (re)constructed. The extent to which this restructuring was part of an active policy or a side effect of spontaneous processes varies. The relation between science/technology and national identity is discussed in a volume edited by Harrison & Johnson (2009). The role of canal-building in specific, with references to Erie Canal is raised as an example of different ways of interpreting transport infrastructure projects; as social, political, cultural or purely financial project. Harrison & Johnson argue that the technological artefacts also have to be taken into consideration when analysing the national identity shaping aspects of transport infrastructure.

National identity is in this relation seen as something that has to do with a unification of an image of the nation-state as both present and primarily a positive value. A more well-defined national identity might thus be at hand, when the general perception of what demarks the nation is widespread and fairly unified among a large share of the population in the geographical area defining a nation. If minorities exist that do not perceive themselves as positively favoured by a more explicit national identity, the majority's concept of national identity can of course be turned into a negative value. Transport infrastructure can of course both link and de-link geographical areas. The image of the national identity might be unevenly spread over a territory. It could perhaps be expected that national identity might be strongest in the nation's centre, such as the capital, while other and competing concepts of national identity might be existing in other more peripheral geographical locations.

National identity might also be seen as something that emanates both from individuals and from collectives. This connects to where the national-identity shaping process originates. If it is part of a central/national policy to shape or reshape the way the nation is perceived, internally or in relation to external actors, it becomes part of a power-exercise, which again brings into play the questions of how the perception of national identity is dispersed in society. Are images of national specificities and demarcations spread in a top-down relation only and do these images allow for alternative interpretations of what can be seen as national identity? Or is it that also decentralized processes expressing national identity are also part of national identity shaping, possibly resulting in a more multifaceted concept of national identity, which possibly does not always adapt to the "central zone's" concept of the nation?

The role of technology and its relation to how national identity and nation-states are defined and constructed is something that also has to be taken into account. It is a recurring situation that large technological systems are often closely connected to central governmental structures for support, politically and financially. In this process what initially starts as a private sector initiative might become included in the national context and the government's actions aiming for furthering a "national plan". By connecting the success of major technological achievements, like Göta kanal or Erie Canal, the value of the nation can be made obvious for the citizens and positive economic or cultural side effects of the transport infrastructure/technology project can be positively connected to the government/state.

Time has already been raised as important in relation to the valuation of major projects' success or role of underpinning national-identity shaping or not. Over time projects like Göta kanal, railway construction or the aviation industry at large might be seen as metaphors for success and future prospect, or as examples of lost opportunities or even worse as technologies threating the sustenance of human life. The valuation of transport infrastructure and transportation might thus be very different seen over time.

As identified for example by Harrison & Johnson, it is easy to end up in a situation where a multiplicity of perspectives and concepts compete when discussing the relation of national identity and social and technological phenomena like transport infrastructure. It seems like the Göta kanal project can be interpreted in many different ways, closely following this multifaceted view.

Göta kanal can be seen as:

- a project proposed by a single entrepreneur who realized the possible benefits of the project, from a commercial and growth-related perspective;
- a project made possible through technological innovation and invention;
- a project that would underpin the formation of a new perception of what could be seen as Swedish as Finland was lost and Sweden had

to reorient its focus to the west, thus supporting the construction of a new national identity;

- a project that supported primarily one specific and rather narrow geographical region and businesses located close to the canal, with the support of substantial state-funding, something that was the starting point for recurring criticism towards the project, and possibly not supporting a "positive" national-identity formation;

- a project the outcome of which varies strongly over time, from representing the technological forefront, to a project being overrun by new technologies (rail) very quickly, but also of military interest and later with importance as a cultural heritage and local growth, and as such with a positive image among many Swedes and supported by politics again, nowadays as a state-owned public corporation (since 1978).

The multifaceted environment that transport infrastructure project like Göta kanal operate within might be summaried as in Fig. 4.5. From a national-identity perspective, a continuous process of construction and contestation of any project is ongoing more or less continuously, from planning/raising support for a project (to the left) to the evaluation of the project ex post (to the right). Projects like transport infrastructure can be used either top down by centralized actors to establish or underpin a national identity or by individuals who further transport infrastructure projects from a bottom-up perspective and try to gain support by arguing based on national-identity-related arguments. Perspectives and challenges change over time, but are more or less always developing.

A possible conclusion here could be that the Göta kanal project had different roles in a national-identity constructing process. Until the final decisions were taken in 1810, it might be seen as a project promoted primarily by a single entrepreneur, primarily Baltzar von Platen, with some support from the central government. As the project was set in motion, combinations of actors and aspects became connected to the project, such as politics, economics and technology, and were used by different actors to promote varying interests over time, but often in a national-identity-related process. The central government at some situations intervened, or had to intervene in order to save the project, and thereby incorporated the project into the central government's priorities, which over time might have fit well or less good with the priorities of the project and its shareholders.

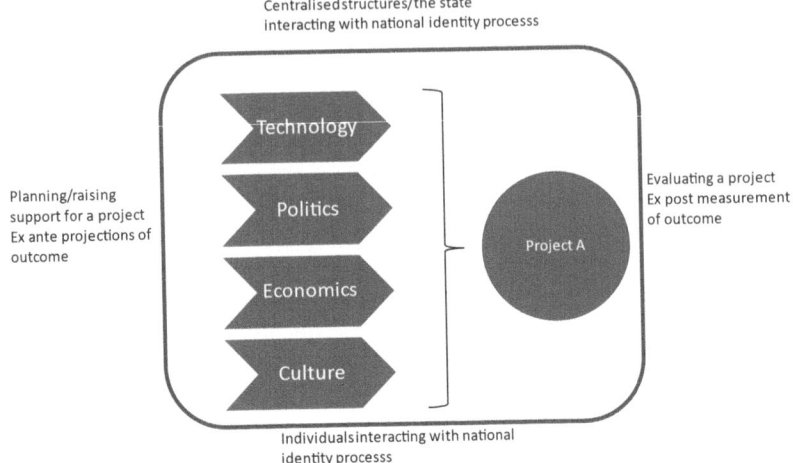

Fig. 4.5   National-Identity shaping and transport infrastructure projects—a model

Following the completion in 1832, additional uses of the canal come more to the fore. Now it was the actual transport flows that were in focus, and the operations (practical and financial) of the canal as a means for achieving many of the aspirations of different actors. A hypothesis could be that the canal was more important as a tool for national-identity formation before and during the construction phase compared to when it was opened for traffic. Transport infrastructure projects like Göta kanal can be seen as involved in and having a role in national-identity processes and can be interpreted in different ways of a project's lifetime, and it is clearly a subject for reinterpretation over time.

TECHNOLOGY IMPORT AND INNOVATION

Canal construction was not new in Sweden at the time the Göta kanal project was started. Also, in earlier projects like Strömsholms kanal and Trollhätte kanal, considerable technological and construction challenges had been met, each according to the level of technology available at the time of construction. It is though obvious that Göta kanal brought technological challenges of a new dimension, taken the large size of the project and the lack of experience in organizing civil works of this size in Sweden at the time.

Locks and bridges had to be manufactured showing the lacking the capacity of the Swedish industry to deliver the necessary construction details in time and referring to the demanded quality was lacking. New materials had to be tested and innovation had to be fostered during the project. Import of knowledge from Britain was, to a large extent, the solution to these challenges.

The willingness to learn from Britain and the many canal projects being carried out there was a sign of the Göta kanal project already from the outset. It seems likely to refer this openness to new production technologies and innovation to von Platen and the chief engineer Samuel Bagge, who both had been involved with Thomas Telford at an early stage of planning the project.

The use of steam engines for pumping water from the excavation sights, the introduction of rail-roads for horse operated railways for transporting stone and gravel and the early introduction of steam-engine machinery in general, and in specific for dredging, are some of the examples of this technology import.

As will be presented in the following chapter the Göta kanal corporation, and in specific von Platen, also went at length in order to establishing a steel and engineering workshop as part of the operation. This possibility had been set out in the Charter of the corporation. Von Platen's actions to establish the workshop, which eventually became Motala verkstad, was not supported by the Board of Directors of the Göta kanal corporation. Initially, therefore, von Platen financed the establishment of the workshop out of own funds. The workshop was eventually established as from 1822, and still is in operation. Over time Motala verkstad has been important for the development of advanced metal engineering and production in Sweden, producing a wide variety of products including steam ships and steam locomotives.

Göta kanal and its workshops with design and engineering of the details of the canal stretches was also an important engineering school for the first generation of civil works engineers in Sweden. Many of the engineers that were trained in the different workplaces along the canal were later to continue with important engineering work, such as the two brothers Nils and John Ericsson. The elder brother Nils would become chief engineer and head of the building project for the first major railway system in Sweden. The younger brother John would become an innovative engineer active in many different fields, but perhaps most well-known for the construction of the warrior ship USS Monitor in the 1860s, which played an important part for the victory of the northern states in the US civil war. Technology import, innovation and development of advanced engineering skills were indeed part of the Göta kanal project.

NOTES

1. Von Platen, B. (1806) *Afhandling om Canaler genom Sverige med särskildt avseende å Wenerns sammanbindande med Östersjön.*
2. Lindgren (1993), p. 115 ff.
3. Von Platen & Telford, T. (1809)
4. More on this in Strömbäck (1993).
5. Bring (1930), del 11 p. 23.
6. Riksdaler—the official currency at the time in Sweden.
7. Von Platen (1806), p. 41.
8. The Bubble Act.
9. Von Platen & Telford, T. (1809).
10. The von Platen letter collection at the Royal Library, Stockholm.
11. See, for example, Magnusson (2022).
12. Charter of the Göta kanal corporation, April 10 1810.
13. Bring (1930, p. 470).
14. p. 238.
15. p. 162 ff, Bring (1930).
16. p. 160, Bring (1930).
17. Other sources claim these were prisoners of war.
18. Von Platen, B. (1822) *Försök till utredning af följderna utaf den arbets-methode vid Götha canal blifvit brukad samt dervid använd kostnad jemte en blick på denna canals blifvande nytta.* Linköping, 1822.

REFERENCES

Björklund, S. (Ed.). (1965). *Kring 1809: om regeringsformens tillkomst* (Vol. 107). ICON Group International.

Bogart, D. (2016, March). Party connections, interest groups and the slow diffusion of infrastructure: Evidence form Britain's first transport revolution. *The*

Economic Journal, 128, 541–575. Royal Economic Society, John Wiley and Sons, UK and USA.

Bring, S. E., et al. (1922/1930). *Göta kanals historia, del 1–2*. Almqvist & Wiksell.

Carlsson, S. (1944). *Gustaf IV Adolfs fall: Krisen i riksstyrelsen, konspirationerna och statsvälvningen (1807–1809)*. Lund.

Harrison, C. E., & Johnson, A. (2009). Introduction: Science and national identity. *Osiris, 24*(1), 1–14. https://doi.org/10.1086/605966

Lindgren, H. (1993). *Kanalbyggarna och staten, offentliga vattenbyggnadsföretag i Sverige från medeltiden till 1810*. Linköping University.

Magnusson, L. (2022). *Från landskapslagar till statsliberalism: det ekonomiska tänkandet i Sverige*. Dialogos Förlag.

Petri, G. (2017). *Hans Järta—en biografi*. Historiska Media.

Strömbäck, L. (1993). *Baltzar von Platen, Thomas Telford och Göta kanal. Entreprenörskap i brytningstid*. Brutus Östlings Symposion.

Trew, A. (2010). Infrastructure finance and industrial takeoff in England. *Journal of Money, Credit and Banking, 42*(6). The Ohio State University.

Von Platen, B. (1806). *Afhandling om Canaler genom Sverige med särskildt avseende å Wenerns sammanbindande med Östersjön*.

Von Platen, B. (1822). *Försök till utredning af följderna utaf den arbets-methode vid Götha Canal blifvit brukad samt dervid använd kostnad jemte en blick på denna canals blifvande nytta*.

Von Platen, B., & Telford, T. (1809). *Berättelser samt kostnads—och ersättnings-förslag rörande den föreslagna Göta kanalen*.

Wodak, R., et al. (2009). *The discursive construction of national identity*. Edinburgh University Press. *ProQuest Ebook Central*, https://ebookcentral.proquest.com/lib/uu/detail.action?docID=1961897

Göta kanal and the Iron Industry in the Adjoining Regions—An Example of Interrelations and Interdependencies

Blif hafvens Salu-Torg
Välmägans nederlag!
- - -
Hell Skandia! Bryt jorden, bryt bergets malm,
Der jernet ligger begrafvet.
Smid Lia till gyllene skördars halm.
Plöj våg genom landet till hafvet.

[Be the marketplace of the seas, the depository of wealth!
- - -
Hail Scandia! Break up the soil, break up the mountain ore, where the iron lies buried.
Forge the scythe for the straw from golden harvests.
Plough the waves through the country to the sea.]
Patriotic song at the completion of the Göta kanal in 1832.[1]

INTRODUCTION

In the spirit of the times, the official opening of the Göta kanal at the small village of Mem on 26 September 1832 was framed by the solemn hymn, a few lines of which introduce this chapter. The hymn placed the new canal in a context of economic and industrial development. It is evident that the celebration was intended to bring out the importance of the canal for the development of traditional industries, not least iron production and agriculture. But the hymn also the potential significance of the canal for the overall development of economy and trade, both within the country and with other countries. Access to new markets (*Salu-Torg* [marketplace]) was to generate additional wealth (*välmågans nederlag* [depository of wealth]).

Up until the beginning of the 1820s, Stjernsund castle[2] south of the nearby town of Askersund in the province of Närke was the centre of one of Sweden's most successful iron mill operations. Even after this period, when the mill enterprises were split up, the iron mill operations in the areas around the manor house remained important economic actors with links to the Göta kanal. Stjernsund and its associated mill operations was one of the places and businesses that were affected by the canal project, both during its construction and after its completion.

The objective of this chapter is to describe the importance of Stjernsund (and its iron mill operations) as a place with connections to the Göta kanal project in its vicinity. The investigation is focused on the relations between three main actors and the connections between Stjernsund and its mill operations and the canal project.

The individuals included in the study are:

- Olof Burenstam (1752–1821), the owner of Stjernsund, an important iron mill owner and a member of the nobility estate at several Riksdag assemblies, e.g., on occasions when the Göta kanal was an important subject of debate.
- Baltzar von Platen (1766–1829), as presented above the leading actor initiating the Göta kanal project, but also a driving force in the realization of the project and in the development of various industries associated with the canal, such as the Motala verkstad workshops.
- Crown Prince/King Karl Johan (1763–1844), who visited Stjernsund on a number of occasions, resided in Burenstam's town property in Örebro during the 1812 Riksdag, and acquired Stjernsund from the heirs of Olof Burenstam in 1822.

Issues investigated include the personal connections between the central actors around Stjernsund as a geographical location, but also Olof

Burenstam's extensive mill operations in a wider sense. In addition, the chapter addresses how the Göta kanal project affected the mill operations in various ways, above all at the Skyllberg mill, just north of Askersund and not far from the Stjernsund manor. To some extent also other parts of Burenstam's operations are discussed (Fig. 5.1).

A further objective is to provide a supplementary perspective on Stjernsund as a geographical location and as the arena for key political and industrial actors during a couple of decades in the early nineteenth century, and to link these to one of Sweden's largest transport infrastructure projects.

The main focus of interest is on the first four decades of the nineteenth century. This period covers the construction of the manor house (completed 1808), Olof Burenstam as a person, owner and leader of the extensive iron mill operations, and the changes due to the transfer of ownership to the royal Bernadotte family in 1822. In all essentials the

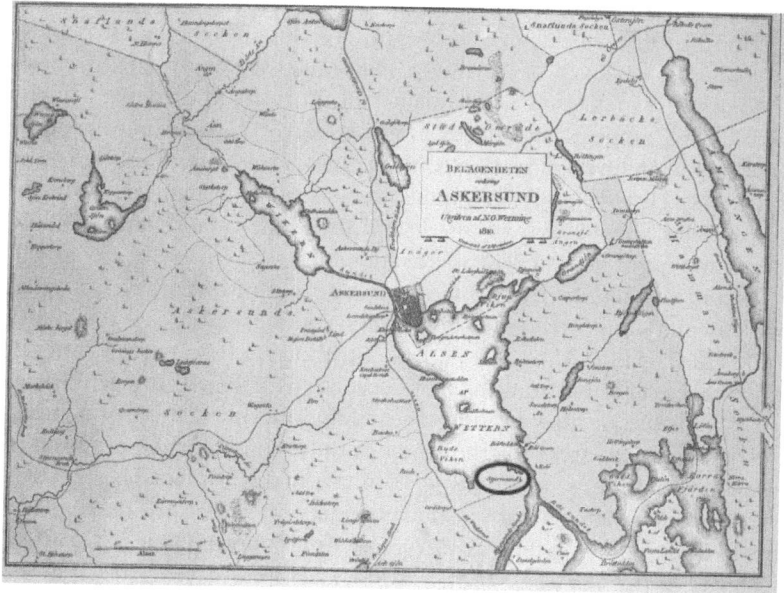

Fig. 5.1 The Askersund region, with Stjernsund south of Askersund (the manor marked) (*Source* The archives of Lantmäteristyrelsen, the National Board of Surveying and Mapping, 1810)

chapter ends, time-wise, by the middle of the 1830s. By then the entire Göta kanal had been inaugurated and the activities within the former canal-building project entered a new phase, with more regular operations.

Sources and Archives

Gustaf Svensson at Skyllbergs Bruks AB, one of the mills forming part of Burenstam's total mill enterprise, kindly and with expertise placed the mill's extensive and well-organized archives at my disposal, which was a great help in searching for material on Stjernsund. For the period when the Bernadotte family were the owners, specifically Karl XIV Johan,[3] access to the Bernadotte archives with the help of Arvid Jakobsson and to the Royal Palace Archives with the help of Mats Hemström, has provided valuable insights and further suggestions for sources and archival searching. The Bernadotte archives should have particular thanks for permission to access the archives for this study.

Searches have been carried out in other archives, including Arkivcentrum in Örebro, where Carl-Magnus Lindgren helpfully advised on searching the archives of Skyllberg mill. The Regional State Archives (Landsarkivet) in Vadstena, the main repository for the archives of the Göta kanal Company, and the National Archives in Stockholm are two other central archives in relation to this study. The library of the Academy of Letters, History and Antiquities, which holds materials related to the Academy's ownership of Stjernsund, is an important source for the understanding of present-day Stjernsund.

Documents held by the AB Göta kanalbolag in Motala have provided supplementary information for my studies. Another resource is the library of the Riksdagen, the Swedish Parliament, allowing repeated searching of printed Riksdag items from the first decades of the nineteenth century, as well as available facilities for digital searching of printed Riksdag items.

Non-archival sources for the study of Stjernsund and the era of Olof Burenstam and subsequent owners, primarily Karl Johan, exist in the form of publications with Stjernsund connections, such as the thematically broad Academy of Letters, History and Antiquities volume, *Stjernsund i Närke—Slottet och godset* [Stjernsund in Närke—The castle and the estate] (2001).[4]

Previous research about Stjernsund has in several ways been important to this study. However, the connections between the Göta kanal and Stjernsund have not previously received explicit attention. The role of Olof Burenstam as mill owner and owner of Stjernsund has been well described by Sven Fritz, both in the above-mentioned 2001 book about Stjernsund and in a separate publication produced by Fritz the following year for the Institute for Economic and Business History Research, Stockholm School of Economics, *Olof Burenstam—Ett bidrag till en brukspatrons biografi* [Olof Burenstam—A contribution to the biography of a mill owner].[5] These studies draw attention to Carolina Camitz (1757–1835). In 1779, she married Olof Burenstam, bringing significant wealth as her portion. After the death of Burenstam, Camitz was responsible for the management of Stjernsund and the mill operations. Occasionally, the source materials provide a glimpse at Carolina Camitz.

Further sources for the study were provided by related economic history research concerning banking and the financial system towards the end of the eighteenth century and the early part of the nineteenth century, i.a. by Sven Fritz. These sources will be noted below where they are cited in the report. In terms of the connections between the Göta kanal and the development of the iron industry, the study provides some additions to the present state of knowledge, providing an account of purchases, i.a. from Burenstam's iron mills. There is an obvious connection here to research on the later expansion of the railway system and its influence on growth. Eli Heckscher's doctoral thesis of 1907[6] and Hans Modig's research[7] on the links between the expansion of the railways and the demand for industrial products are two such examples.

Olof Burenstam's brother Carl Daniel Burén, the owner of the Boxholm mill also situated at Lake Vättern, during a large part of his life kept a voluminous diary covering many different issues of the time, including political and economic questions. The diaries have been transcribed and recently published by the National Library of Sweden,[8] together with an introduction to assist in reading the diaries.[9] The diaries provide a certain amount of supplementary information about the period and about Stjernsund, as well as about his brother Olof's ownership of the castle and the mills. Carl-Gustaf Burén has kindly assisted with searches of the voluminous material for the aspects studied here.

A key source regarding the Skyllberg mill and the period when Burenstam was its owner is the book written for the 300-year anniversary of Skyllbergs Bruks AB in 1946, *Skyllberg 1346*1646*1946*, of which part 2, for the period 1775–1946 is most relevant here.[10]

Karl Johan has been portrayed in several books. The biography most frequently used for this study is Torsten Höjer's classic biography *Carl XIV Johan* in three volumes, published 1939–1960. Here it is primarily parts 2 and 3 that are of interest.

Stjernsund and its associated mill operations are relevant to an analysis of the Göta kanal project in terms of several of the analytical dimensions applied in this book. The Stjernsund estate was established as a manifestation of the success experienced by the owner. The iron and steel industry and its various branches played and does still play an important part in the Swedish economy. The extraction and processing of iron ore allowed access to important basic materials for various uses in the community and for the industrialization that began during the period under review. The iron industry also generated considerable added value and became an important export sector in eighteenth–nineteenth centuries Sweden.

Within the framework of existing regulation of the iron industry, with its strict rules about production and state regulations intervening in operations in several different areas, it was possible for successful and entrepreneurial actors to create important business operations, in some cases also generating substantial private wealth. As the creator of Stjernsund, Burenstam was one of the more successful of these actors and also one whose ambition was to achieve an established role in society. The Stjernsund manor house was probably intended as part of his demonstration of achievement and success.

Burenstam owned a number of iron mills and factories in Närke and Östergötland, both provinces close to Göta kanal. The iron mill operations were in all essentials carried out on site in the different production locations. The operations of Burenstam's iron mills and the surplus these generated were the basis for the creation of the Stjernsund manor house, and it is this that provides a more obvious link to the Göta kanal. The mill operations created a business flow with various financial transactions as an important part, and they also generated a need for transport which could be satisfied through various modes of transport, among which maritime transport was an efficient alternative.

The products of the mills, both bar iron and iron manufactures, were also important inputs to the Göta kanal project. The Göta kanal operations and the traffic that started in 1822 on the western stretch (from Rödesund at Lake Vättern to Sjötorp at Läke Vänern and further southbound to Gothenburg) eventually came to be of considerable importance for the Skyllberg mill, the part of Burenstam's mill operations that is the main focus of this study.

There are a number of personal links between Stjernsund and the Göta kanal. The individuals included in the study were introduced above. Olof Burenstam was a successful mill owner, both during the Gustavian era of the late eighteenth and nineteenth centuries and under the new regime from 1809. Burenstam's solid position in public life also outside the iron mill environment is evidenced not only by his participation in the 1812 Riksdag (as well as those of 1800 and 1809–1810), but also by providing his Örebro property as residence for the new Crown Prince, Karl Johan, for the Örebro Riksdag. The purchase of Stjernsund by the Bernadotte family after the death of Burenstam, and their retention of the estate for about 40 years, might well be a result of Karl Johan forming connections with Stjernsund and this part of the country already in the early stages of his Swedish life.

The Göta kanal was, for its time, a very large project. This had an extensive impact on society and created economic activity at a level previously unknown in Sweden. It was followed by another large infrastructure project, the railways, which also greatly influenced society and in its implementation in several respects followed the pattern established by the Göta kanal project. Operations of such importance to the economy as the mill industry owned by Olof Burenstam were affected by this large project in several ways. Here a broad overview of the history of Stjernsund is the starting point.

The Stjernsund Manor and the Iron Mill Business Operation

The estate Stjernsund was purchased by Olof Burenstam in 1785, then with a manor house dating back to the seventeenth century. The property was agricultural, with a relatively large number of tenant farms and cottages and, significantly, the Dohnafors iron mill situated to the west of Stjernsund (Fig. 5.2).

It can be assumed that it was the iron mill that provided the impetus for the acquisition, as it could be added to the growing mill operations built up by Burenstam from the late 1770s onwards. The Dohnafors mill was joined operationally to the Skyllberg mill, which served as a central point for the mills located in the province Närke. Burenstam's total mill

Fig. 5.2   Stjernsund castle (*Photo* Björn Hasselgren)

business also included farms and iron mill operations in Östergötland, chiefly at Grytgöl and Sonstorp.

As Fritz (2002) has pointed out, it was probably the lively market for iron during the 1790s that made it possible for Burenstam to begin the extensive manor house building project. Burenstam moreover had experience of and evidently interest in building projects of the kind represented by Stjernsund. The manor house at Skyllberg saw relatively extensive additions and reconstructions once Burenstam had acquired it in 1775. Burenstam also organized extensive reconstruction projects at Sonstorp, one of the Östergötland mills.

When Burenstam appears in the rather sparse sources concerning him as a person that are cited in the literature and found in the archives, it is as an active owner of an iron mill business. The then current regulation of the iron industry meant that the principal means of extending iron mill operations was by acquiring other iron mills and farms with forge privileges. The size of the production of forged or bar iron was the subject of close state regulation, monitored by the Bergskollegium authority (Swedish Board of Mines).[11] This way of adding existing mills to his operations was also the means used by Burenstam during several of his active decades. Fritz compares the total production at Burenstam's mills, at c. 4500–5000 *skeppund*[12] of bar iron, to the largest mills in Sweden, Lövsta and Uddeholm, which produced 6000–7000 *skeppund* annually.

Burenstam appears in several respects to have had an early interest in progressing his social and business career. Unlike his brother Carl Daniel Burén, the owner of Boxholm mill, Olof Burenstam sought a place in the nobility. His objective was, as he expressed it in his May 1788 application to Gustav III to be rewarded by entry into the nobility, "*from life's uncertainties to seek the consolation, when I depart from my beloved children, to be able through this precious seal to confirm the assurances I have given them about the rewards of virtue and merit*".[13] The application was granted in 1791 and, as was the practice, intended to be implemented a year later in 1792, which was then delayed to 1793 following the murder of Gustav III.

Burenstam had relatively close contacts with key layers of Swedish society. During a period in the early nineteenth century, he had, for instance, close contact by letter with Gustaf Mauritz Armfelt, then commanding the Swedish forces in Pomerania, and previously a close ally of Gustav III. Burenstam's son, Johan Daniel Burenstam, was a cavalry captain and Armfelt's aide-de-camp. A number of letters from this period are preserved in the Swedish National Archives, which exemplifies a relatively close connection between Burenstam and Armfelt.[14]

Waldén (1949) also notes as mentioned earlier that Crown Prince Karl Johan resided in Burenstam's property in Örebro during the Örebro Riksdag of 1812, and a document in the (Stockholm) Royal Palace Archives provides evidence of a visit by Karl Johan to Stjernsund in 1815 (see below).

By the time of his death in 1821, Burenstam was registered as living at Stjernsund after living many years at his other estates, also after the completion of the Stjernsund manor. The Stjernsund manor house had finally become the residence of one of Sweden's most successful iron mill owners. However, it would appear that Stjernsund never developed into a distinct and long-term centre for the iron mill business.

Olof Burenstam and Stjernsund manor house were, as has been noted above, well known to Karl Johan at the time of the death of Burenstam in 1821. At the distribution of Burenstam's estate, his assets including the mill business were divided between his widow and his children. The totality of the Stjernsund property was split between the mill business and the manor house and its surrounding grounds as separate lots before the sale in 1822.

In this connection, it might be noted that Otto Julius Hagelstam, the husband of one of Burenstam's daughters, was a hydrographer and a naval officer. He was involved with marine cartography over a large number of years, i.a. participating in the preparations for the Göta kanal in 1808.[15] Here, as well, we thus find a link between the Burenstam family and the Göta kanal project. Hagelstam was also interested in the organization of the navy and appears to have held views on its organization that were close to those espoused by Baltzar von Platen, with a significant role for the archipelago fleet (Skärgårdsflottan—the fleet of the Army), whose ships were considered in determining the dimensions of the Göta kanal.

During "the autumn" of 1822, Karl Johan and Queen Desideria and members of their court are said to have visited Stjernsund for three days, after which, according to a written account[16] found in the Stjernsund manor archives, the company left for Norway, at the time in union with Sweden. It seems reasonable to assume that this visit coincided with the inauguration of the western stretch of the Göta kanal, from Rödesund to Sjötorp. The inauguration took place on 23 September 1822. The account in the Stjernsund manor archives, stated to have been written based on information from Fritz Burenstam, who at the time of the visit was only ten years old, should of course be treated with some caution as a source. In terms of its timing, the visit does however appear likely.

Bring's "*Notiser om Västgötalinjens tillkomst och invigning*" [Notes on the creation and inauguration of the Västergötland stretch],[17] which forms part of the extensive Göta kanal books, includes a statement, not, however, supported by source references, that Karl Johan met von Platen at Stjernsund, ahead of the inauguration of the western stretch of the Göta kanal. Karl Johan left Stockholm on 17 September 1822, according to Bring. The journey went via Norrköping, Linköping, Motala and Medevi, a route that does not seem to have been the usual one when the court travelled west to Norway, according to what can be gleaned from information in the archives of the Royal Court Stables (Hovstallet). After the inauguration of the western stretch of the Göta kanal on 24 September, the King continued his journey "to Christiania" (the name of the Norwegian capital at the time, as stated by Bring).

According to the official account of the inauguration of the Västergöt-land stretch of the Göta kanal on 23–24 September 1822,[18] Karl Johan's visit to the canal lasted from the evening of 22 September to the morning of 25 September. This document includes, in its introduction, precisely the statement that Bring most likely refers to, namely that Karl Johan, before the visit, was staying at Stjernsund and that he was there "*waited upon by the former Member of the Council of Ministers and Admiral, Count von Platen*".[19]

As was noted above, the sources mentioning this visit by Karl Johan to Stjernsund, one of the few occasions when it is possible to link several of the actors at the centre of this study to this location at the same time, are relatively uncertain. A dinner is supposed to have been given at Stjernsund in the course of Karl Johan's stay there, to which the Skyllberg family and others were said to have been invited. It can reasonably be assumed that the question of the purchase of Stjernsund, or even the conclusion of the business, was part of the dinner conversation. According to the purchase agreement, written as early as 11 September 1822, Karl Johan finally took possession of Stjernsund on 14 March 1823.[20]

GÖTA KANAL AND STJERNSUND, WITH THE IRON MILL OPERATIONS—IN RELATION TO THE ANALYTICAL FRAMEWORK IN THIS BOOK

During the period under review, Stjernsund became something of an arena for business interests, general development interests in the region of northern Lake Vättern, and overall, for more explicitly political interests. These different spheres of interest touch upon the different aspects of technology, economics and politics which form the basis of the analytical framework of this book.

Göta kanal as a Political Project

Like Olof Burenstam, von Platen succeeded in repositioning himself at the time of the shift from the previous Gustavian regime to the new regime established after the revolution of 1809. The political system established by the new constitution, with a balance of power between ruler, Riksdag and courts, provided a framework for the continued development of society and the economy with partly new rules of the game.

At the time of the transformation of 1809, Burenstam had been an established member of the nobility for 15 years and had, for example, participated in the 1800, 1809–1810 och 1812 Riksdag meetings.[21] Burenstam was a member of the Committee on appeals and the economy at the 1809–1810 Riksdag and of the Committee on taxation and tolls at the 1812 Riksdag, principally speaking on issues concerning the conditions for the operation of iron manufacture and mill businesses. One example is a memorandum presented by Burenstam in the proceedings of the nobility estate on 1 December 1809, arguing in favour of the existing arrangements within the pig iron trade and its regulation against a deregulation of the pig iron trade,[22] arrangements that were fundamental to Burenstam's iron mill operations. Burenstam had obviously an interest in preserving regulations that were hindering a more liberal development of trade and commerce in those cases when his business would be exposed to less favourable conditions by a deregulation.

The Göta kanal issue was one of the major items for discussion at those Riksdag meetings where Burenstam participated from 1809. The issue was discussed to such an extent that it is unreasonable to believe that Burenstam would not have been well informed about the different aspects of the project. Burenstam was wealthy and the owner of one of the country's largest industrial enterprises; he had close contacts with trade interests in both Stockholm and Gothenburg through the sale of his produce. Transportation costs were among the larger at the time in the iron mill business, the improvement of transportation would be a natural political issue to get involved in with this background.

A compilation produced by Carl-Gustaf Burén[23] also shows that questions related to the canal were a recurrent theme in Carl Daniel Burén's diaries. Like his brother, Olof Burenstam, he operated an extensive mill business. The diaries contain just under 50 annotations about general canal issues, some of which touch on the Göta kanal, often about the progress of the projects and comparisons between different canal projects. Canals as infrastructure projects were undoubtedly relevant to the owners of iron mill businesses during this period even if Olof Burenstam was not part of the recorded Riksdag debates.

The Göta kanal Project—Making Use of Technology, With and Against the Iron Mill Businesses

An obvious cause for concern for the Göta kanal corporation, as described above, was the substantial deficit in engineering knowledge as well as in the ability and production capacity to manufacture significant construction components for the canal, not least in iron and steel. For example, there was a need for a considerable number of cast or moulded constructions, not least for bridges and lock gates. A recurring problem was that the availability of pig iron for such work was limited, but also that the quality of those casts produced in Sweden was inadequate. The Göta kanal corporation tried to manage this situation by placing orders with those iron mills that were deemed to have sufficient production capacity and knowledge and also through imports from Britain. This, however, was subject to restrictive regulations, which made the import procedures cumbersome. Here one might of course note Olof Burenstam's support for the existing regulation of the availability of pig iron at the 1809 Riksdag, intended as protection for the traditional export of bar iron. This did not make things easier for a large project like the Göta kanal.

Burenstam's iron mill operations collected in the business attached to Stjernsund can be seen as an aspect of those external conditions that—based on the institutional framework in existence in the early nineteenth century—created the preconditions for the implementation of a project like the Göta kanal. The iron industry was the subject of detailed regulation both as regards iron ore mining and the production's subsequent processing stages, and these regulations were gradually reduced only when the canal project was already under way.

The primary idea behind the regulation of the iron industry, where the technological development was slow, was to concentrate on production of bar iron for export, by converting pig iron into bar iron which could then be sold abroad. It was such activities that Burenstam's operations, for instance in Skyllberg, focused on. The way the Stjernsund iron mills were operated, thus, did not always facilitate the Göta kanal project.

At the same time, Fritz[24] points out that a relatively large proportion of the products of the totality of businesses run by Burenstam at the various mills around Stjernsund consisted of finished iron manufactures such as nails, iron sheets, saw blades for water-driven sawmills and anchors. In particular, it can be seen that this type of production increased from 1810 onwards, when the Bergskollegium agreed an increase of the output of the forge at Sonstorp, one of Burenstam's mills.[25] These products were of the kinds that were in demand for the canal project, where there was a need for a large number of many different manufactured goods. Furthermore, it is possible to find evidence that Burenstam's mills were the suppliers of such goods to the canal project.

This can be seen in scattered annotations in the Göta kanal archives, from as early as 1810,[26] that indicate that of Burenstam's mills Sonstorp, supplied relatively large amounts of "iron nails", "iron and nails", "materials and objects" (e.g., hinges) to the canal project. While the orders were not particularly large, seen in relation to Burenstam's total business operations, they nevertheless illustrate that the Göta kanal project had an impact on Burenstam's businesses also in this respect.

The Göta kanal corporation's so-called Order books (Orderböcker), preserved in the County Archives in Vadstena, also contain the record of a "procurement auction" held in the regional administrative city adjacent to the western stretch Mariestad on 16 January 1812[27] (or at least recorded that day). The record details a considerable amount of manufactured iron goods bought for the canal project from several of the mills around Lake Vättern, such as Forsvik mill, Boxholm mill (run by Burenstam's brother, Burén) and Påvelstorp. Products include "breakers", "cleavers", square iron of various dimensions, "root axes", "nail punches" and nails in large quantities and of several dimensions. While arguing for the preservation of the regulations safe guarding the supply of pig iron to the iron mills, and thus making the demand from the Göta kanal more difficult to meet in the domestic market for some goods, Burenstam in reality seems to have exploited the possibilities to increase sales of manufacture, when the regulations on that part of the iron business were lifted.

One of the options discussed for dealing with the supply problems encountered by the Göta kanal project in relation to more advanced workshop products, such as cast-iron lock gates and bridges, was as mentioned above the possible establishment of in-house manufacture of these products. Such a possibility had been included in the charter for the canal of 1810.[28] At von Platen's instigation, concrete discussions had been started as early as 1817 about establishing workshop operations within the project. The solution to this challenge within the project was once more provided through contacts with Britain and Thomas Telford. An English foundry master, Thomson, was employed and he recommended the establishment of a foundry and workshop by Lake Viken, in a—now insignificant—place called Sörkvarn, south of the Sätra mill along the

western canal stretch. Arguments in favour of locating a workshop installation, there were for example, the easy access to iron ore from areas like
the Lerbäck mining region, particularly in the vicinity of Stjernsund and
Burenstam's mill businesses.[29]

After several complicated developments within the Göta kanal project
and following the financial crisis in the corporation during 1817–1818 the
workshop and foundry ended up being established in Motala instead of
in Sörkvarn. One reason might simply have been that the western stretch
of the Göta kanal had by then been completed and was open to traffic.
The construction work and the need for cast iron goods of various kinds
were instead concentrated to the eastern part of the canal and for this, the
growing town Motala was the natural hub. This part of region Östergötland, close to the Närke mining regions, also had a lengthy history as
one of the most important southern centres of Swedish iron production.
There was therefore a certain degree of familiarity with iron production and forge work in the region. However, a more detailed study of
recruitment patterns for the Motala verkstad workshops covering a somewhat later period (1853–1860), cited by Dahlström (1998)[30] indicates
that during this period, the majority of workers were recruited from the
surrounding rural areas, and not from the iron industry or from Motala.

It remains to establish the more exact total extent of supplies to the
canal project from Burenstam's mills. It is likely that there was a significant increase in demand for bar iron and manufactured iron goods in
the regions along the canal created by the canal project. Burenstam and
other mill owners in the region responded to this demand. It is possible
that the establishment of the Motala verkstad workshops in 1822 eventually meant that mills already established in this market experienced greater
competition. But this takes us in the main into the 1830s and outside of
the period primarily addressed by the present study.

A similar attempt to investigate links between infrastructure and industrial development, in this case in connection with the major railway
projects undertaken in Sweden during the period 1860–1914, can be
found in Hans Modig's doctoral dissertation of 1971.[31] Based on
an extensive review of orders from the public railway projects and the
private railway companies, Modig concluded that the railway companies
were certainly major customers of the Swedish iron mills and workshop
companies. At the same time, Modig estimates that the totality of railway
orders did not amount to more than two to three per cent of the collected
output of the "metal industry"[32] for these years. For the "workshop

sector", the highest proportion noted was 22.5% in 1875, naturally of great importance for a single year, but at the same time not a level that could be sustained long term.

Modig describes the situation of the railway companies as in many respects similar to that of the Göta kanal corporation. In addition to deficiencies in general domestic production capacity, it was also noted that the quality of the supplied products was inadequate. This also applied to the Motala verkstad workshops who had difficulty supplying, for example, railway tracks at the same price and quality as was done by the foreign, mainly English, suppliers.

The experience gained at the beginning of the nineteenth century, when Olof Burenstam's operations experienced the new level of demand from a big project like the Göta kanal, demonstrates the significance of such a project for the total demand for manufactured iron goods, but primarily for less advanced goods. However, it also shows that import of the more advanced product groups became a pattern that was established at an early stage and remained valid for subsequent decades. Only slowly was a domestic production capacity for more advanced iron goods developed. All told, the relatively sparse information about the demand from the Göta kanal project suggests that there was an operational and commercial link between the Göta kanal and the Stjernsund mills, but that this most likely did not amount to something of decisive operational or economic significance for the mills. The iron mill business at the time of the Göta kanal project was still characterised by the remaining regulation, leading to a focus on export of bar iron.

The Göta kanal as an Economic and Financial Actor, Influencing the Conditions of Industry

The Göta kanal project also had an impact on the broader economic and financial developments along the canal, including for example, the financial interactions between actors in the Närke province, and grew to play an essential part for significant actors during the first decades of the nineteenth century. The project had an impact on economic development in several ways. The construction project as such led to large numbers of interactions between the Göta kanal corporation and other economic actors for the duration of the construction period. This might be questions of land issues, purchase of various inputs to the building and for the project's employees and workers, examples of which have been given above, but also purely financial aspects that will be further touched upon below. Furthermore, the completed project reduced transport and travel times and opened up new markets for the regions' products.

The charter of the corporation initially states, as a kind of goal for the whole project and the Göta kanal corporation, that it is to lead to *"relief for agriculture and useful industries, lively trade and mobility and constant and increasing diligence, and furthermore become the main conduit for several possible future branch canals through the most fertile provinces in the country"*.[33] These effects might arise not only during the period of construction, but also after the completion of the canal, since transport in the regions associated with the canal would be influenced and new and faster transport relations established. There is reason to believe that both effects were of interest to the State as co-financer of the canal, but also to those private stakeholders who helped provide finance by subscribing to shares in the canal corporation.

The canal corporation was of special interest to the two regions, Östergötland and Västergötland, traversed by the canal, something that is evidenced by the fact that the whole share issue prospectus, including the charter and von Platen's and Telford's investigations, was read out in the churches in these counties at the morning service on 6 May 1810. This might not have been unique; many public notifications and proclamations were disseminated in this way, but the project nevertheless had the character of something like a regional mobilization effort.

In terms of the financial aspects, it should, as already mentioned, be noted that the Göta kanal corporation charter permitted the establishment of the Göta kanal Diskont (*Göta kanaldiskonten*). It was a sister bank to two similar businesses, the Malmö Discount House and the Gothenburg Discount House, already established. The entities in Malmö and Gothenburg were created in 1803 and 1802, respectively, with the objective of facilitating the provision of credit for private industry and trade, and this was also the intention in respect of the Göta kanal Diskont. These discount houses had been preceded by similar operations towards the end of the eighteenth century, the Discount Company (*Diskontkompaniet*) in Stockholm and the Gothenburg Discount Office (*Göteborgs Diskontkontor*) in Gothenburg, with regulation and functions similar to the discount houses of the early nineteenth century. The earlier discount houses had been liquidated in 1788 and 1795.

The new discount houses, owned by private interests, operated under time limited licences. In the case of the Göta kanal Diskont, the intended period of operations was primarily linked to the financing of the construction project, with a stated terminal date at 25 years after the beginning of the project, i.e., 1835. The discount houses accepted deposits and provided loans in the same way as modern banks. They could also issue a kind of banknotes, deposit certificates (discount notes) and assignations which were a way of utilizing an approved credit by way of a promissory note that could be deployed in financial transactions, at a time when there was often a pronounced shortage of banknotes in circulation. The operations of the discount houses were supported by access to credit in the Central Bank (the Riksens ständers bank), but their lending and general operations were restricted by several operational regulations.

The Göta kanal Diskont developed into an important financial actor in those parts of the country where the canal was built but also in a broader regional market, with activities both on the deposit side and as a provider of credit. It is also likely that the various notes that the Diskont was able to issue were important to the general trade.

As far as Burenstam and the mill enterprises are concerned, it may be noted that there is no evidence that Burenstam was a subscriber to the initial share issue in 1810, at least not in his own name. There would, however, have been good opportunities for him to participate in the share subscription, with facilities available in three locations close to Stjernsund, Sonstorp and Skyllberg, namely Örebro, Vadstena and Linköping. A review of the preserved share subscription lists[34] does not provide any indications that Olof Burenstam might have participated in the subscription. There is one person, Samuel Troilius,[35] who was involved with the mill operations in Skyllberg and who, according to Fritz, might have acted as Burenstam's deputy at the mill. Troilius subscribed, in his own name, to a small number of shares at the share subscription in Örebro in May 1810.

The share subscription lists for Vadstena include Olof Burenstam's brother, Carl Daniel Burén, and one further person belonging to the Burén family, against relatively large amounts. It might be surprising that Olof Burenstam does not appear to have been interested in participating as an owner in a corporation like the Göta kanal, with potential to play such an important role for the mill enterprises. One explanation might be that the project appeared risky and that Burenstam was unwilling, or unable, to increase his financial exposure further after ventures such as the costly construction of the manor house at Stjernsund.

However, an inventory dated 1821[36] of the assets of the estate of the late Olof Burenstam lists shares in the Göta kanal corporation amounting to 10.000 Rdr, even though these were said to be pledged as security for loans from the Göta kanal Diskont, something that was common among the share subscribers. Shares in Trollhätte and Hjälmare lock companies, both of smaller amounts, were listed in the estate inventory. It is not clear how or when these shares were subscribed to or became the property of Olof Burenstam, but it nevertheless indicates that financial instruments related to the Göta kanal came to be introduced into Burenstam's business affairs.

Burenstam's enterprises were also in touch with the Göta kanal Diskont in other contexts. Fritz (2002) points to several different links between Burenstam and the Diskont. On the one hand, it concerned credit provided for the ordinary activities within the mill enterprise, with credit of various kinds regularly used in the sale of the output of bar iron through Stockholm or Gothenburg as export locations, the dominant ones for the Skyllberg operations. On the other hand, there are a number of other borrowing transactions involving the Diskont and where Burenstam, or his heirs, were listed as borrowers during the period after his death. Fritz states that the inventory of Olof Burenstam's estate in the spring of 1821 included a debt to the Diskont of 8.490 Rdr, representing about 8 per cent of the balance sheet in respect of the businesses left by Burenstam. Credit continued to be given also after the death of Burenstam, as his widow, Carolina Camitz, signed a debt instrument for example in April 1821 in favour of the Diskont.

Even though the Diskont ended up being liquidated from 1818, due to a financial crisis in the Malmö Discount House, which led to the State revoking the operating licences for all three discount houses, it can still be seen that the Diskont continued to have a role in the financing of industry during a relatively long period after its active operations. The liquidation was handled by the Swedish National Debt Office (Riksgäldskontoret) and went on for a long time. Andersson (1983) states that the liquidation was completed in 1848 and only after extensive involvement by the Riksbank (*Riksens ständers bank*).

A further aspect of the impact of the Diskont on the activities in Burenstam's mill enterprises was that it might have meant that the choice of export ports for bar iron varied in line with the availability of credit. Fritz (2002) mentions that a large part of the production of the mills in Närke went via Gothenburg during the period of greatest activity for the Diskont, after which it reverted to being directed to Stockholm, a transport route going from Skyllberg via Örebro, north of Askersund and Skyllberg. The availability of credit and liquidity might therefore have played a part in the business decisions, and the Göta kanal Diskont influenced these to a great extent for a couple of decades at the beginning of the nineteenth century.

An interesting detail in this context, referred to by Fritz, is that correspondence from 1812 about a loan transaction in the Diskont indicates a direct contact between Olof Burenstam and G. H. Ekman, a member of the board of the Diskont but also the brother-in-law of Baltzar von Platen and moreover active in the Göta kanal corporation, for example, as an important shareholder. It is also in itself a reflection of the very limited size of the highest levels of society in nineteenth-century Sweden.

To sum up, it can be seen that there were several different connections between the Göta kanal project and Burenstam's business operations, both in his lifetime and after 1821. The Göta kanal corporation and the Göta kanal Diskont soon developed into important economic actors not only in the regions where the canal was constructed but also in a broader sense for the financial system, by offering deposit and credit facilities as well as payment methods through the various types of banknotes issued.

Provision of credit appears to have had a place in the more short-term business flow in the form of operating credits, which the iron mill business was continually in need of due to the way operations were dominated by the changing seasons. Costs were large in the early part of the year and income later on in the year when it was possible to sell processed iron. But various credits through the Diskont were also important for the more long-term financing of the operations, for example, in connection with purchases of property and businesses.

Göta kanal Alters Geographical Connections and Transport Time

One of the principal objectives of the construction of the Göta kanal was to connect the North Sea with the Baltic Sea and in this way create new possibilities for both passenger travel and transport of goods. As mentioned above, the emphasis was on broadly trade and business facilitating effects, and these were most obvious in relation to transport of goods.

The Göta kanal fitted well into this larger strategical perspective. Later on in the project, these overarching strategical arguments were strengthened through the union with Norway of 1814. The King's opening address at the 1815 Riksdag emphasized the successes of the Göta kanal project in spite of its many difficulties, but also the increased importance of the project due to the union with Norway:

> ...this enterprise that, due to later events, has seen a manifold increase in importance in that the union of the two seas has become more important due to the union of the two kingdoms....[37]

An important line in the argumentation in favour of the Göta kanal centred on transport costs. Here the possible comparators were road transport or maritime transport following the coastline between coastal towns. For transport over greater distances, between the North Sea and the Baltic Sea via Öresund, Öresund customs duties had to be paid to Denmark. Waterborne transport through the country could offer greater load capacity than road transport, shorter transport distances compared to coastal maritime transport and overall savings in terms of the use of horses and other draft animals. An example from the 1815 Riksdag debates illustrates how the gains, expressed in the number of draft animals used, were presented by the representatives of the canal, in this case by von Platen.

Thus, von Platen stated in 1815, according to Bring (1930),[38] that transport by road of the volume of traffic anticipated on the canal would require 400,000 "man days" and 800,000 "horse days", compared with the calculation for the canal of 39,000 and 3,000 such "days" respectively. Such greater efficiency was expected to be able to generate savings in the use of agricultural land for growing food for people and animals, to the extent of 16,087 *tunnland* (or about 8,000 hectares), from which should be subtracted 587 *tunnland* to be used by the canal. It was therefore likely that construction of a canal would generate macro-economic as well as business-economic efficiencies. Provided one chooses to believe von Platen's line of argument.

The argument against canals as a means of transport in a country like Sweden with a pronounced winter season, was of course that the canal could not be expected to remain open during the winter months when ice formation made operation impossible. The Göta kanal could be expected to remain open during eight, at most nine, months of the year. For the rest of the year, transport had to be over land, which after all meant that such transport would be used during the part of the year best suited to it, during the period of deep ground frost.

It was specifically the possibility of all-year-round transport that was one of the clearest reasons for the strong position of the railways a couple of decades after the inauguration of the Göta kanal. In addition, the railways provided a further increase in the speed of transport, compared to the canal traffic. While the latter was faster than the road option for the transport of heavy goods, it was relatively slow, using (initially) sailing ships which for long stretches of canal had to be pulled by men or draft animals.

Only with the introduction of steam ships in the 1830s and 1840s was it possible for canal traffic to achieve shorter transport times. One piece of information[39] about travel times states that a passenger voyage by steam ship between Stockholm and Gothenburg via the Göta kanal took about 56 hours in the 1870s, while the corresponding journey by train took about 14 hours. It can be assumed that transport of goods on the Göta kanal, before steam ships became more common, took significantly more time than the 56 hours stated here.

From the later patterns of use of the Göta kanal, it is possible to conclude that the canal traffic probably became most important over time for heavy long-distance transports. Statistics over goods transported on the canal indicate[40] that during the early years of the 1830s, the product groups "Pig iron and scrap iron" and "Iron and steel, processed" together formed the dominant types of product transported on the canal. Gradually, grain, wood products, stone and other heavy product groups tended to increase in volume, while processed iron and steel products decreased. It is likely that these transports gradually moved over to transport by rail.

For the mill enterprises belonging to—or previously belonging to—Burenstam's business group, the transport question was undoubtedly decisive for the control of the overall cost levels within bar iron production. It is likely that the Göta kanal, above all, affected the transport conditions for those mills that were associated with Lake Vättern, for which a speedier transport route for export via Gothenburg was now able to compete with export via Stockholm.

Fritz (2002) has observed that the totality of transport connected with the mill enterprise in Skyllberg, was as complicated as for other iron mills. The production of charcoal was the single largest task where transport was included, though the charcoal transport was restricted to the local area. The raw materials for the production of bar iron were pig iron from

Nora (some 80 km north of Skyllberg) or from the neighbouring mining operations in the Lerbäck area (north of Askersund, towards Örebro). The ready made bar iron was sent via Örebro for further transport via Stockholm, or via Askersund (at least once the western stretch had been opened in 1822) to Gothenburg.

An account[41] concerning operations at Skyllberg mill during 1823–1832, said to have been compiled probably in the 1830s, by Johan Daniel Burenstam (Olof Burenstam's son), provides certain basic facts in terms of the organization and costs of transport for the mill:

- The distance to Stockholm and Gothenburg, respectively (the export ports) is given as 25 *mil* (250 km), including **10 km** by road to Askersund (for Gothenburg) or **45 km** to Örebro (for Stockholm).
- Transport cost for pig iron (from Örebro to Skyllberg) amounts to 20–25 *skillingar Riksgälds* per *skeppund* in summer and 28–32 *skillingar* in winter.
- The transport cost for bar iron amounts to 1 *Rdr* and ½ *skilling banco* for transport via Askersund and for transport from Örebro to Stockholm to "about" 28 *skillingar banco* (just above ½ *Rdr*).

These data suggest that the transport cost from Skyllberg via Örebro to Stockholm or via Askersund to Gothenburg would be approximately the same, if it is assumed that the road transport cost for the pig iron would be similar to the road transport cost in the opposite direction for bar iron to Örebro. The saving arising from choosing the route via Askersund and the shorter road transport stretch may have been outweighed by the higher transport cost by ship from Askersund to Gothenburg than from Örebro to Stockholm.

As Fritz points out, financing possibilities and transport costs were both included in the total calculation that determined the choice of transport route for bar iron for export. This means that iron from Skyllberg alternated between being exported through Örebro/Stockholm and Gothenburg during the first decades of the nineteenth century. At the time of Burenstam's death in 1821, the transports had been going via the Stockholm route during a couple of years. At this stage of Burenstam's operations, the choice of export port may have been determined by which trading establishment offered the most advantageous trade credit for the mill enterprises and at the time that was Stockholm. But this was before the western section of the Göta kanal was opened.

A review of parts of the materials in the mill archives at Skyllberg mill for the early decades of the nineteenth century confirms the picture given by Fritz. Consignments from Skyllberg alternate between going via Örebro and onwards to Stockholm and via Askersund and onwards to (Uddevalla or) Gothenburg. Further on, in the early 1830s, the impression is that shipping via Askersund has become dominant. It is possible to follow individual transports via Askersund in the company accounts[42] and to compare these with the sailing lists preserved in the archives of the Göta kanal corporation.[43] Here one can also find timings of transports from Askersund via Lake Vättern and onwards via the western stretch of the Göta kanal to Sjötorp by Lake Vänern, all carefully recorded by ship, transport date and time of entry and exit of the locks.

One example is that of the ship *Carolus*, marked down for a load in "Journals regarding bar iron for Gothenburg" in the accounts of the Skyllberg mill in Askersund on 29 April 1833, arrived at the Forsvik lock (the first one in the Göta kanal from Lake Vättern in the direction of Lake Vänern) at 08:00 on 9 May 1833 and left the Sjötorp lock station (the last one in the direction of Lake Vänern) at 06:00 on 11 May 1833. Thus, this portion of the Skyllberg to Gothenburg transport took 46 hours, which suggests relatively slow transport compared with steam ships and, in particular, with the railways a couple of decades later. Compared to land transport to Örebro, it is reasonable to assume that canal transport was the faster alternative before steam ships were introduced.

The size of the load for this particular transport is entered in the sailing lists of the Göta kanal as 581 *skeppund*,[44] or about 79 tonnes. The weight entered in the Skyllberg accounts is lower than that stated in the sailing details from the Göta kanal locks. This might be because the ship in question also loaded bar iron from other mills. At the locks it was stated that bar iron was the greater part of the cargo carried by the ship Carolus, information that determined the class of fees charged for the vessel.

The same ship, and a number of others, are recorded as having carried out several transports southbound or northbound from and to Askersund during the year, in the transport relations described above. To sum up, there was a well-developed transport system in connection with the Skyllberg mill for the iron transports of the 1830s, and the Göta kanal had an obvious part in it. Previously, the transports had more often alternated between Stockholm and Gothenburg but here we can see a transport route, which most likely provided the most efficient transport for the bar iron (possibly also in financial terms), during a period before the establishment of the railways.

Concluding Comments

This objective of this chapter is to throw light on a number of aspects of Stjernsund, its manor and the associated iron mill enterprise, as a place where various events related to the Göta kanal project unfolded. Several sources can confirm that during the period under examination, 1800 to the 1830s, the Göta kanal project affected Olof Burenstam's business enterprise, but also that some of the central actors in the events surrounding the Göta kanal were attached to, or had relations with, Stjernsund in various ways.

As regards personal connections evidenced by various sources, a number of individuals have been tracked in this analysis: Olof Burenstam, Baltzar von Platen and Crown Prince/King Karl Johan.

With the exception of one or two instances, it is not possible to find evidence in the sources that this study is based on that the individuals tracked here actually got together at the manor house at Stjernsund. At the same time, it appears likely that Burenstam, von Platen and Karl Johan all visited the Stjernsund manor and had good knowledge of each other, and it is not entirely out of the question that they actually got together at Stjernsund. In one way this can be seen as natural, given the limited extent of the Swedish social elite at the time. There were close connections between the nobility and the royal family.

It is somewhat easier to find evidence of links between the mill enterprises at Olof Burenstam's various mills and the Göta kanal. Here the connection to Stjernsund as a place is less direct. It has to do with how the Göta kanal changed the conditions for transport, which can to some extent be illustrated by the study of transport routes for the Skyllberg mill described above, where the creation of the canal provided better transport facilities for the products of the mill to the Gothenburg port for onward shipment. Another aspect is that the Göta kanal Corporation was most likely a repeat customer for other parts of Burenstam's mill enterprise, primarily Sonstorp.

As a financial actor, the Göta kanal was of great significance through the banking operations carried out by the Göta kanal Diskont, but also because many individuals living and working in the regions near the canal invested in shares in the corporation. Relative to previous studies, it has been possible to add a couple of further aspects of these financial connections to the picture of Olof Burenstam's extensive mill enterprise, which was finally—at least in terms of ownership—centred on the Stjernsund manor where Burenstam and his family eventually made their home. The picture of the mill enterprise and Burenstam's finances as intertwined with the canal project in several different ways, also in purely financial terms—as shareholder, borrower and user of the various different means of payment provided by the Diskont—has thus been added to.

All told, the study links Olof Burenstam's mill operations with the Göta kanal project in a more elaborate way than previously demonstrated. There were personal connections, business connections and financial connections between the different actors in the Östergötland and Närke regions around Stjernsund. The Göta kanal was a major infrastructure project that affected the conditions for other industrial operations and trade.

Thus, the Göta kanal did, in several different ways, stimulate the economic activity in the regions where it was constructed and brought into use, and this included Stjernsund. In this way, the canal project did, in some respects, fulfil the aims of providing "*a help for agriculture and useful industries, promotion of trade and transport, and uninterrupted and increasing industriousness*". Olof Burenstam formed part of this development, in terms of industry as well as of politics, although, as it seems, his perspective appears primarily to have been that of the mill owner rather than that of the politician.

Acknowledgements The research behind this chapter has been facilitated by financial support from the Royal Swedish Academy of Letters, History and Antiquities (Kungliga Vitterhetsakademien), Dr Carl Kempe's Foundation.

Notes

1. Pontin (1832).
2. The Stjernsund estate is generally referred to as a "castle", but this is not unproblematic. It would be more correct to regard Stjernsund as a manor house built to be the principal manor on the Stjernsund estate. The label "castle" has been applied to Stjernsund since the nineteenth century and is currently the more or less official designation of the property. It might be seen from the subsequent text that contemporary observers on

the one hand described Stjernsund as a "house" or "small castle" (the architect Carl Fredrik Sundvall) or "stone buildings" (Carl-Daniel Burén, the brother of Olof Burenstam). The present study mostly uses the label "Stjernsund castle" or "Stjernsund manor house" when the reference is to the manor house building, and "Stjernsund" when the reference is to the estate in a wider sense, including the mill operations.

3. Henceforth "Karl Johan" and sometimes "the Crown Prince" or "the King" are used to refer to Karl XIV Johan.

4. Åman (2001).

5. Fritz (2002).

6. Heckscher (1907).

7. Modig (1971).

8. Burén (2019b).

9. Burén (2019a).

10. Waldén (1949).

11. For an overview of iron industry regulation and the activities of the Bergskollegium, see Isacson (1997).

12. 1 *skeppund* was between 136 and 195 kilograms depending on the unit measured.

13. Translated quotation from Olof Burén's application for ennoblement, published in Waldén (1949).

14. See further below, Sect. 4.2.

15. Otto Julius Hagelstam, urn:sbl:13,498, *Svenskt biografiskt lexikon* (article by Herman Richter), accessed 2022-09-15.

16. Stjernsund Archives Vol. 001, APP1d, Cassel family, 1834–1905.

17. Bring (1922).

18. [Account of the opening of the Göta kanal in Västergötland in the august presence of His Majesty the King] Berättelse angående öppnandet av Götha Canal i Westergötland under Hans Maj:ts Konungens egen höga närvaro (1822).

19. p. 1.

20. Stockholm. Bernadottearkivet, Kungens enskilda domäner (Egendoms-förvaltningen i Syd-, Väst- och Mellansverige), vol. 69 Kontrakt och handlingar gällande Stjärnsund 1822–1830.

21. See i.a. Seth (1935).

22. Olof Burenstam's memorandum, reproduced in Riksdagens Protokoll 1 december 1809 från sammanträde med Allmänna Besvärs- och Oeconomie utskottet [The Protocol of the Riksdag, 1 December 1809, of a meeting of the General Appeals and Economy Committee].

23. Compilation kindly done at my request, from Carl Daniel Burén's diary.

24. In Åman (2001).

25. See Fritz (2002, p. 43).

26. Vouchers 1810 for Göta kanal and Berg station etc., vouchers 1 May, 26 July and 15 October. Vadstena. Landsarkivet. Göta kanal-arkivet, G XI a:1 [Göta kanal Archives].

27. Order book 1812. Vadstena. Landsarkivet. Göta kanalbolags arkiv, B IV a:1 [Göta kanal Archives].

28. Section 7 of the Charter of the Göta kanal Company, dated 11 April 1810, gave the company the following rights: "for the requirements of the canal, its construction and maintenance, to establish workshops for the manufacture of cast iron goods, forged goods, tools and utensils, and various necessary materials, and to power these workshops either manually or using water or steam engines...".

29. Lerbäck can be seen northwest of Askersund in Fig. 5.1.

30. Dahlström (1998) citing Hoppe (1988).

31. Modig (1971, pp. 95 ff).

32. A concept derived from the national accounts for Sweden calculated by Lindahl, Dahlgren & Kock (1937) and including the subsectors "mechanical engineering" and "metal industry", which covers the majority of the businesses in this broader sector during the period investigated by Modig.

33. Introduction to the Royal decision (oktroj) to establish the Göta kanal corporation (1810).

34. Vadstena. Landsarkivet. Göta kanal-arkivet [Göta kanal Archives], Subscription lists (G II f).

35. Fritz (2002, p. 46).

36. Skyllberg. Skyllbergs bruksarkiv [Skyllberg Mill Archives] Bruksägaresläkter [Mill owning families] (F 2:1) – Sterbhusets skulder april 1821.

37. [Account of events since the previous Riksdag]. "*Detta företag, som igenom senare inträffade händelser, vunnit en mångdubblad betydenhet, i det att föreningen af båda hafven, blef genom båda Rikenas förening, af en ökad vigt*". Berättelse om hvad sig i Rikets Styrelse tilldragit, sedan sista Riksdag; Gifven Stockholms Slott à Rikssalen, den 6 Mars 1815, s 146, BIHANG Till samteliga Riks-Ståndens Protokoll.

38. This section of Bring (1930, part 1:3 p. 219), was written by Herbert Lund, as noted in the book.

39. AB Göta kanalbolag, website, 2022-07-17 Passagerartrafiken (gotakanal.se).

40. Tersmeden in Bring (1930, vol. II, pp. 184–185).

41. Waldén (1949, pp. 396 ff), one *Rdr* (*Rdr*) was divided into 48 *skilling banco* – *sk:bco*; one *rdr Riksgälds* (*rgs*) was divided into 24 *sk rgs*; one *rdr bco* = 1.5 *sk rgs*.

42. Skyllberg. Skyllbergs bruksarkiv [Skyllberg Mill Archives] G2 BA:11–12 for the years 1832–1833, and G3 BA for 1833.

43. Vadstena. Landsarkivet. Göta kanalbolags arkiv [Göta kanal Archives], e.g. GV aa 30 Seglationsspecial för år 1833.
44. 1 *skeppund stapelstadsvikt* = 136 kg, one of the different measures mentioned in foot note 12.

References

Archives

Örebro. Folkrörelsearkivet [Social Movement Archives]
Vadstena. Landsarkivet. Göta kanalbolags arkiv [Göta kanal Corporation Archives]
Skyllberg. Skyllbergs bruksarkiv [Skyllberg Mill Archives]
Stjernsund. Slottsarkivet [Archives of the Stjernsund castle]
Stockholm. Bernadotte-arkivet, Stockholms Slott [Bernadotte Family Archives, Royal Palace]
Stockholm. Riksarkivet [Swedish National Archives]
Stockholm. Riksarkivet/Slottsarkivet [Royal Palace Archives]
Stockholm. Riksarkivet. Sjöholmsarkivet [Sjöholm Archives]
Stockholm. Vitterhetsakademiens bibliotek [The Library of the Royal Swedish Academy of Letters, History and Antiquities]

Government Publications

[Account of the opening of the Göta kanal in Västergötland in the august presence of His Majesty the King] Berättelse angående öppnandet af Götha Canal i Westergöthland under hans Maj:ts Konungens egen höga närvaro, tryckt hos F.J. Lewerentz's Enka, Mariæstad, 1822
[Olof Burenstam's memorandum reproduced in 'Riksdagens Protokoll' 1 December 1809 in a meeting of the General Appeals and Economy Committee.] Riksdagens Protokoll 1 december 1809 från sammanträde med Allmänna Besvärs- och Oeconomie utskottet
[Account of events since the previous Riksdag]. Bihang Till samteliga Riks-Ståndens Protocoll, Innehållande de Handlingar, Hvilka till Riks-Ståndens blifvit aflemnade, Såsom Konungens Nådiga Propositioner, Riksens Ständers Utskotts Betänkanden och Förklaringar, Samt Riks-Ståndens Protokolls - Utdrag, vid Urtima Riksdagen i Stockholm 1815, Stockholm 1815
[Charter of the Göta kanal Corporation, 1810] Privilegium för Götha Canal Bolag, givet Stockholms Slott den 11 April 1810

[Regulations for the Göta kanal Corporation, 11 April 1810] Reglor för det för det genom Kungl Majts nådiga privilegium af den 11 April 1810 Octroyerade Götha Canal Bolag Gifne Stockholms Slott den 11 April 1810

LITERATURE

Andersson, B. (1983). Early history of banking in Gothenburg discount house operations 1783–1818. *Scandinavian Economic History Review, 31*(1), 50–67.

Bring, S. E., et al. (1922/1930). *Göta kanals historia, del 1–2*. Almqvist & Wiksell.

Burén, C. G. (2019a). *Brukspatron i brytningstid. En introduktion till Carl Daniel Buréns dagböcker 1790–1815* (Acta 92:1). Kungliga biblioteket.

Burén, C. G. (2019b). *Brukspatron i brytningstid. En fullständig renskrift av Carl Daniel Buréns dagböcker 1790–1815* (Acta 92:2). Kungliga biblioteket.

Dahlström, E. (1998). Verkstaden vid kanalen, Motala verkstad under 1800-talet. *Daedalus, 66*, 1–22.

Fritz, S. (1990). Commercial Banking in late Eighteenth Century Sweden. In E. Cieślak & O. Henryk (Eds.), *Changes in two Baltic countries: Poland and Sweden in the XVIIIth century*. No. 164. UAM.

Fritz, S. (2002). *Olof Burenstam, Ett bidrag till en brukspatrons biografi*. Forskningsrapport 15, Institutet för ekonomiskhistorisk forskning. Handelshögskolan.

Heckscher, E. F. (1907). *Till belysning af järnvägarnas betydelse för Sveriges ekonomiska utveckling*. Centraltryckeriet.

Hoppe, G. (1988). Från jordbruk till järnverk: Kopplingar mellan jordbruk, förindustriell massproduktion och formell industri ur ett geografiskt perspektiv med exempel från 1850-talets Motala. *Bebyggelsehistorisk Tidskrift, 16*, 77–92.

Höjer, T. T. (1960). *Carl XIV Johan, Konungatiden*. Norstedt.

Isacson, M. (1997). Bergskollegium och den tidigindustriella järnhanteringen. In *Daedalus: Tekniska Museets årsbok 1998, Människa, teknik, industri* (pp. 43–58). Tekniska museet. Stockholm.

Karlsson, P. A. (1990). *Järnbruken och ståndssamhället: institutionell och attitydmässig konflikt under Sveriges tidiga industrialisering 1700–1770*. Doctoral dissertation, Stockholm University.

Modig, H. (1971). *Järnvägarnas efterfrågan och den svenska industrin 1860–1914*. Norstedt.

Pontin, M. M. (1832). *Fosterländsk Sång vid Götha Canals fullbordan*. Norstedt.

Seth, M. (1935). *Personregister till rikets ständers protokoll för tiden från och med år 1809 till och med år 1866*. Riksgäldskontoret. http://urn.kb.se/resolve?urn=urn:nbn:se:kb:riks-21798908

Svenskt Biografiskt Lexikon, information about several individuals mentioned in the report.

Waldén, B. (1949). *Skyllberg 1346*1646*1946: minnesskrift på uppdrag av Styrelsen för Skyllbergs Bruks Aktiebolag, Andra delen, tiden 1775–1946.*

Åman, A. (Ed.). (2001). *Stjernsund i Närke, Slottet och godset,* Kungliga Vitterhets Historie- och Antikvitets-akademien. Almqvist och Wicksell.

CHAPTER 6

Conclusions and Analysis

The Göta kanal project was a great achievement in many ways in early 1800s Sweden. In a situation where Sweden had lost most of its overseas territories and was struck by the financial restraints caused both by the loss of Finland, the war with Russia and conflicts in Europe, Sweden was also in upheaval over its constitution and the structure of the central government. Even if most of these challenges were dealt with without too strong divided views, conflicts and turmoil, the combined effects for Sweden and for the government at large were, of course, very serious. It stands out as quite exceptional that the Göta kanal project, even if it had been initiated before the 1809–1810 events, could be decided on and established in the midst of these challenging times.

One of the main reasons explaining that the decisions were taken in this situation discussed in this book has been that the project clearly had a function to contribute to the unification of the country around a new challenging endeavour, something that could restore the national pride and support the formation of a new national identity. But, of course, the project was also used as a tool in order to refocus trade and transport more to the west than earlier. The national symbolic value of the great canal project was emphasized by the leading nineteenth-century poet Esaias Tegnér in his 1811 poem "Svea",[1] where a direct allusion to the canal project was framed in the most nationalistic and powerful language in this verse:

© The Author(s), under exclusive license to Springer Nature Switzerland AG 2023
B. Hasselgren, *An Institutional Approach to the Göta kanal*,
https://doi.org/10.1007/978-3-031-44416-6_6

Lead the waves of the river around as tamed subjects,
and within Sweden's borders conquer Finland again!

The connection between the canal project, national identity formation and Tegnér's poem has been pointed at, among others by Malmborg (2001). In Chapter 4, the role of transport infrastructure projects like Göta kanal as regards national identity formation has also been discussed.

Following the project from initiation to completion, it is obvious that it can be seen either as having fulfilled its goals to provide a new and more effective sailing-waterway where a strong vision for the reduction of the transportation distance between east and west in Sweden was successfully addressed. To further trade and industriousness, as were among the other goals of the project expressed in the Privilege, is more difficult to evaluate and has been discussed over time.

Seen from another angle Göta kanal is an example of a project for which the project-planning and design phases were unusually unsuccessful. Costs soared compared to the first estimate to such an extent that it seems questionable if the original estimates were at all based on sufficient knowledge of the challenges ahead.

The time frames for the finalization of the project were also changed to an extent that is rarely seen in transport infrastructure projects (though Sweden and other countries have experienced such examples also in recent years). The long project time, which spans over three decades, also (naturally) led to a gradual shift in focus on the arguments that were emphasized in the external communication of the Göta kanal corporation in relation to the many debates about the project. It also led to successive alterations as to which of the aspects that influenced the project could be seen as more or less important.

Going back to the analytical framework in this paper in Chapter 2, the different arguments and influential aspects over time might be structured as in Table 6.1.

The time line of the entire canal project in Table 6.1 is divided into three periods: "the initiation/planning phase", the first ten years of construction ("the independence phase") and the remaining 13 years until finalization ("the national business project"). Over time there was a continuous political and public debate and discourse involving most of the different arguments and areas of influence classified in Table 6.1 as the main areas driving the development. The focus of the process was (often)

Table 6.1 The development of the Göta kanal project, arguments used over time

Time period	Technology	Economics	Politics	Dominant organizational principle
1800–1809: "Initiation/ planning phase"	Modern canals had successfully been constructed in other countries and this should be done in Sweden as well	Benefits for the public economy form part of the reasoning. Transport costs to be reduced	Linking eastern and western Sweden, growth and increased trade, aspects of the liberalization of the economy, a national identity shaping effect	Organizational issues not discussed
1810–1818: "The independence phase"	The company management gradually come to understand the project's great technological challenges and the lack of technological and engineering knowledge and skills in Sweden. The connection to Britain is strengthened	Rapid completion of the project is sought since it will generate revenue. Another way of adding to the income is to expand the Göta kanal Diskont. When the project receives criticism, for instance in the 1815 Riksdag, its representatives revert to the benefit arguments listed above	Initially, there is strong emphasis on the company's independence from the State. The arguments in favour of state involvement gradually grow stronger. Defence policy links, as the basis for these arguments, grow stronger over time	From an independent private sector initiative and project to a public/ private collaborative project

(continued)

Table 6.1 (continued)

Time period	Technology	Economics	Politics	Dominant organizational principle
1819–1832: "The national business project"	Technological development and education of engineers and workers. Motala verkstad as an important part of the project and for technological education and innovation generally	Focus on finding sufficient finance for the completion of the project within a short-term perspective. Focus on opening the entire canal so that operational revenue would be generated	Arguments in favour of the long-term vision behind the project are increasing. Stronger focus on innovation, growth and defence in the argumentation	A state-controlled, privately owned limited company

the formal role and status of the corporation as regards ownership and financing.

From a focus on independence, as the project had been decided in 1809–1810, the soaring construction costs and crisis of the Göta kanal Diskont in 1817 led to a change of the basic financial arrangement, in that the project became dependent on continuous financial support from the government. This was also an event that made Göta kanal to become a consistent issue in the parliamentary sessions from 1817 to 1830.

Technology as a driving force for development seems to have played different roles as arguments as well. During the initiation phase, as argumentation in favour of the project was the main focus, it seems as if the availability of appropriate technology and engineering skill in order to actually build the canal was more or less taken for granted. There were experiences from other canal projects in Sweden and also a knowledge based on projects in other countries. As the project was established and construction started, the lack of appropriate technologies and skill was, however, soon experienced by the corporation, and in ever new areas. The solution to attract and acquire knowledge from Britain by forming the collaboration with Thomas Telford and different associates to Telford promised to solve some of these deficiencies. But these issues continued to be among the main concerns of the project until the finalization. Establishing the metal workshop Motala verkstad as one solution to these problems has been discussed above.

Socio economic benefits and financial aspects of the project were recurring in the argumentation in relation to the project. As for economics there was a strong line of argumentation based on the wider social-economic benefits the canal was argued to bring, primarily used in the argumentation during the pre-1809 phase. A number of examples of how the economy would thrive from the canal was produced and reiterated in the debates in Parliament and in the public domain. As the project was established these arguments tended to be of less prominence in the discussion, but were raised again in different ways in relation to the many occasions the project was put into question.

Critique related to deficiencies as regards project management, the operations of the Diskont or soaring costs were all countered by the corporation with arguments in relation to the wealth and growth bringing propensities of the canal. These were arguments to a large extent based in a liberal view on the economy at large and with a positive view on the effects of increased trade and commerce. Here the leading proponents of the project, such as Järta, Skogman and von Platen illustrated a knowledge of how economics was analysed and discussed in Britain and other countries leading the industrial revolution. Adam Smith was among the scholars who explicitly inspired this generation, and both Järta and Skogman were active in introducing Smith's ideas in the Swedish discussion (see Petri, 2017).

As the project was successively finalized in different stretches there was also a strong focus in the project on the expansion of the transportation flows on the canal. This would show the positive sides of the project to the public and the opposition in Parliament and it would also generate revenue for the corporation. Over time the focus of the management of the canal corporation shifted more and more to this operational perspective on the canal. The study of the actual use of the canal for the adjacent iron industry around Lake Vättern in a previous chapter has illustrated that the project as such had a role for the metal works in the region but the canal also for the transport flows of bar iron to Gothenburg and the export markets.

Politics was over time continuously an area and basis of argumentation in the debate around the canal. Some different perspectives in the political regime of the time were brought to the fore and exemplified during the different phases and in relation to the changing political environment. One important such phase was the initiation of the project, during the last years of the older supreme rule system with the King Gustav IV. Here, perhaps the decision to initiate the project might be seen as part of the aspirations for a general modernization of the economy that, at least according to Carlsson (1944), characterized the regime until the upheaval in 1809. A sign of the underlying solid support for the Göta kanal project was the fact that the plans survived the restoration and even might have become stronger from the loss of Finland in 1809.

The fact that von Platen, formerly something of an outsider in national politics, suddenly became one of the important political actors with a seat in the new cabinet and a role in the Constitutional Committee of the Parliament, was indeed something unexpected. An explanation might be the close connections von Platen had in the years before the 1809 restoration with one of the leading figures in the revolutionary process Adlersparre, but perhaps also with the new king Karl XIII. Adlersparre and von Platen had discussed the need for a change in the leadership of the country and von Platen had sent his 1806 report with a plan for the Göta kanal to Adlersparre, asking for comments.

Both of them were also enthusiastic about strengthening the connections to Norway in different ways. Both had important roles in the decision to elect the Danish prince Karl-August, governor of Norway, to become the new Swedish crown prince in 1810. After leaving the cabinet in 1812, von Platen kept close connections to Karl XIII and the new Crown Prince (following the unexpected death of Karl August in 1810), Karl Johan.

Von Platen remained one of the strongest supporters of the new regime, even at times when Karl Johan seemed to take a role far stronger than the stipulated in the 1809 constitution. Von Platen also was one of the strong supporters of the union with Norway and held important roles in relation to Norway, e.g., as Swedish governor in Oslo. Von Platen's close connections to the central government of course were very valuable for the corporation also after he had left the cabinet. Even if von Platen and Karl XIV Johan had some serious conflicts during the canal project, primarily in relation to the decision to close the Diskont in 1817, the King though continued to support von Platen and the canal.

Different policy areas were involved in the political argumentation in relation to the canal project. As mentioned above, wealth and growth policies in general was one area of discussion. Trade and commerce policies was another subject often under discussion in relation to the canal. And over time there was a growing interest in the defence policies of the country as a new defence strategy, the "Central Location Defence-strategy" (Centralförsvarstanken), was developed. According to this strategy, defence resources should be centred around Lake Vättern with a fortification and heavy military investments in Karlsborg, a small town on the western side of Lake Vättern, having been made more accessible through Göta kanal. The need for this military strategy to be developed and the possibility for Göta kanal to have a role in that development was often mentioned, during the later stages of the project, by von Platen. The actual importance of these arguments for the continuation of the project is though difficult to assess. As a matter of fact, the Karlsborg fortress was constructed as one of the major defence assets in Sweden.

Industrial policies at large were also part of the argumentation in relation to the canal. The establishment of the Motala verkstad as an offspring of the project was the theme of debates in Parliament and was a source for criticism towards the project. It can either be seen as a foresighted investment bringing Sweden into the coming industrialization era, or as an example of the mis-management of the canal project at large.

The Göta kanal project was planned and constructed in a situation where previous canal projects had been carried out both in Sweden and in other countries. Most well researched is the connection between the contemporary canal project in Scotland, the Caledonian Canal, where a direct relationship existed based on the contacts between Baltzar von Platen and Thomas Telford. Here both technology and general knowledge as regards many of the aspects of canal construction were imported to Sweden, both in theory and in practice, as engineers and workers from Britain came to Sweden to work on the project. Another important aspect that unites the Caledonian Canal and Göta kanal was, according to von Platen (1822), the dimension of the canal, which would in both cases allow for larger (also military) ships, an important aspect for von Platen.

When it comes to financing, the Caledonian Canal and Göta kanal are less similar. The Caledonian Canal was a more clear-cut state-financed project, with less involvement from the private sector. The Erie Canal, which in many ways shared the general idea of regional development and increased availability with Göta kanal, was also a great achievement as regards technology and construction, where the Göta kanal project probably could have learned more than what seems to have been the case. From a financial point of view, the Erie Canal was also clearly part of the public sector, though with the State of New York as the borrower in the financial market.

There seems to have been closer logical connections between some of the other canal projects in Britain than with the Caledonian Canal. The analysis of financial aspects that Ward presented in his 1974 investigation shows similarities with the organization of Göta kanal, with a stronger private sector initiative than in Scotland and in New York State. Similarities can also be found with the French canal projects 1814–1848, where the involvement of private sector actors was more developed than for the Caledonian and the Erie canals. In his 1822 report on the construction methods and costs, von Platen (mentioned above) referred to many of the canals in Sweden and in other countries, including France. It though seems like von Platen was not too well informed on the contemporary canal programme in France, which could have inspired as much as the older canal projects in other countries that von Platen mentioned. Clearly, there was a wide field of knowledge and experiences to draw upon for the actors involved in the Göta kanal project. While the connections to Britain have been described, other connections and inspirations could be further analysed.

The way of analysing the Göta kanal project, which has been the basis of the presentation in this book, shows the possibility to structure the vast documentation of the project in different phases under which different challenges and arguments were more influential than other. Discussions and disputes, but also support for the canal, are possible to describe in a condensed way and to connect to the continuous efforts to define the project in relation to the government and to the private financers and formal owners of the corporation.

Communication from von Platen and other central actors in the corporation over time was influential for how the project was perceived. Different patterns of co-evolution are also possible to see. Over time the different aspects of argumentation seem to have been combined to support the project, sometimes in new patterns. In this way, the co-evolutionary and institutional analytical framework brings new light to the Göta kanal project, which makes it more easily analysed and compared to other similar projects.

NOTE

1. 204 (Esaias Tegnérs Samlade skrifter/Andra bandet) (runeberg.org).

REFERENCES

LITERATURE

Carlsson, S. (1944). *Gustaf IV Adolfs fall: krisen i riksstyrelsen, konspirationerna och statsvälvningen (1807–1809)*.
Malmborg, M. A. (2001). *Neutrality and state building in Sweden*. Springer.
Petri, G. (2017). *Hans Järta—en biografi*. Historiska Media.
Von Platen, B. (1822). *Försök till utredning af följderna utaf den arbets-methode vid Götha canal blifvit brukad samt dervid använd kostnad jemte en blick på denna canals blifvande nytta*.
Ward, J. R. (1974). *The finance of canal building in eighteenth century England*. Oxford University Press.

INDEX

GPSR Compliance

The European Union's (EU) General Product Safety Regulation (GPSR) is a set of rules that requires consumer products to be safe and our obligations to ensure this.

If you have any concerns about our products, you can contact us on ProductSafety@springernature.com

In case Publisher is established outside the EU, the EU authorized representative is:

Springer Nature Customer Service Center GmbH
Europaplatz 3
69115 Heidelberg, Germany

The manufacturer's authorised representative in the EU is Springer
Nature Customer Service Centre GmbH, Europaplatz 3, 69115 Heidelberg,
Germany. If you have any concerns regarding our products, please
contact ProductSafety@springernature.com

Printed and bound by CPI Group (UK) Ltd, Croydon, CR0 4YY

29/04/2026

02099546-0001